INTERFERENCE AVOIDANCE
METHODS FOR
WIRELESS SYSTEMS

Information Technology: Transmission, Processing, and Storage

A Continuation Order Plan is available for this series. A continuation order will bring delivery of each new volume
immediately upon publication. Volumes are billed only upon actual shipment. For further information please contact
the publisher.

INTERFERENCE AVOIDANCE METHODS FOR WIRELESS SYSTEMS

Dimitrie C. Popescu

University of Texas at San Antonio
San Antonio, Texas

and

Christopher Rose

Rutgers University
New Brunswick, New Jersey

Kluwer Academic / Plenum Publishers
NEW YORK, BOSTON, DORDRECHT, LONDON, MOSCOW

Library of Congress Cataloging-in-Publication Data

ISBN: 0-306-48188-X

©2004 Kluwer Academic/Plenum Publishers, New York
233 Spring Street, New York, New York 10013

http://www.kluweronline.com

10 9 8 7 6 5 4 3 2 1

A C.I.P. record for this book is available from the Library of Congress

Permissions for books published in Europe: *permissions@wkap.nl*
Permissions for books published in the United States of America: *permissions@wkap.com*

Printed in the United States of America

Preface

Wireless systems use radio which has an uncanny ability to flow through and around obstacles, pervading space and providing tetherless connectivity to users on the move. However, this marvelous feature also leads to mutual interference between different systems and mutual interference, in general, severely degrades performance. Our work grew from investigations of mutual interference and methods of counteracting it.

Historically, the earliest method of interference control was to license discrete non-overlapping swaths of radio spectrum to different systems. Of course, such policies hail from the days when building a transmission facility required various feats of technical cunning and brilliance along with large capital expenditure. That is, the technical difficulty associated with delivering commercial radio or television in the earlier part of the 20th century required large investment of funds and therefore large stable returns on investment. Since mutual interference could make systems inoperable, legal structures were erected to afford tight regulation of seemingly scarce spectrum resources. By and large, this somewhat archaic structure – based on older technologies and a perceived scarcity of radio resources – persists to this day.

However, though licensing has led to financially successful and ubiquitous wireless services such as radio, television and cellular phones, the fear of mutual interference and the associated mentality of licensing can have stultifying effects on grass roots innovation of services – even when such services could eventually become equally ubiquitous and successful.

For example, consider recent 802.11 (WiFi for short) systems. These local wireless data networks are springing up everywhere because they offer inexpensive tetherless access to data in the home, in the workplace, and increasingly in public spaces such as malls, parks and other congregation points. A recent study in New York City showed that there are thousands of privately owned access points scattered throughout the lower part of Manhattan, most ostensibly for office and personal use, though many are installed in coffee shops and the like.

One could imagine that such an installed base might be attractive to a larger telecommunications concern which could form a coalition which cobbled these access points into a larger network allowing nearly ubiquitous access. Since voice can be carried over the Internet using VOIP services, one could also imagine a form of low cost "cellular" service riding atop such WiFi networks – all for a fee to the larger service provider. Such musings (conveniently ignoring certain technical details) beg the question of how cellular telephony could compete against such seemingly low cost distributed services.

The major problem with this scenario is control and quality-of-service provision owing to mutual interference that cannot be controlled in the unlicensed bands occupied by current WiFi systems or effectively mitigated using their current technical methods. For instance, a small number of strategically placed and completely legal transceivers (rogues, or simply other systems) could severely disrupt the operation of the larger system. With no technical way to combat interference and no legal cease and desist recourse open to the large coalition, the only tactic available would amount to bribery of the rogues – an unstable model at best. Thus, it seems unlikely that we will see ubiquitous for-fee access to wireless data over WiFi networks anytime soon and the licensed structure of cellular for its concomitantly large investment seems eminently reasonable.

However, suppose WiFi transceivers could naturally coordinate their use of spectral resources to minimize interference? Then, the effect of rogues, even persistent and malicious ones, might be minimized. Furthermore, if the rogues actually comprise a different service whose primary purpose is not disruption, it would be in their interests to coordinate spectrum use with other systems resident in their spectrum and perhaps some measure of peaceful coexistence could be achieved.

Such notions might have been pipe dreams in the past, but with the advent of agile radios which can change their methods of modulation on the fly, such musings could begin to approach engineering reality. Thus, the assumption of agile radios was our intellectual entry point and immediately prompted a question: what signal processing/transmission methods could different systems employ to allow peaceful coexistence?

One answer, built on the notion of ubiquitous Internet access, might be to have "spectrum servers" which helped users coordinate their local spectrum use. However, we felt methods which required direct collaboration between disparate users and systems might be unworkable (as might the assumption of ubiquitous access to the Internet). So we sought "distributed" methods which could be used independent of what other users employed such as interference suppression (using MMSE filters, for instance). However, it became immediately apparent that if the radio is agile, one could also change the modulation format and from this notion sprang idea of interference avoidance considered in this monograph where users, following only self interest, adjust their modulation to maximize their use of the available spectrum.

This monograph takes the reader carefully through our (and others') early discoveries regarding interference avoidance and seeks to promote the notion of a shared signal space over which users can rove when thinking about problems of mutual interference. The theory is developed first for single users using single codewords at a single receiver over simple wireless channels and gradually extended to more general wireless scenarios including fading channels and general multiaccess vector channel systems.

Now all that said, we must also admit to an analytic failure. Though we have suspicions from simulations and some hand-waving arguments based on analysis for older cellular systems (CDMA in particular), we do not yet have answers to even the simplest problems where different systems mutually interfere. That is, we consider primarily mutual interference between different users of a given system. Why? One answer is that the network information theory problems engendered by such mutual interference problems (the interference channel, the relay channel and the like) have eluded solution for over fifty years. Thus, we have no real means of comparing the performance of interference avoidance to an accepted metric as we do in the single system case.

Nonetheless, we and others are working hard on this problem and some simpler, practical variants which use interference avoidance to coordinate spectrum sharing in a transparent and efficient manner. We of course refer the interested reader to this recent work and perhaps in the due course of time, these results will form the basis for a second volume.

October 2003

DIMITRIE C. POPESCU

CHRISTOPHER ROSE

Contents

Chapter 1

INTRODUCTION

Interference from natural sources and from other users of the medium has always bedeviled the design of reliable communication systems. In wired communication systems the amount of interference can be limited by restricting physical access to the medium and/or reducing relative noise energy. However, no such restrictions can be applied in wireless systems where the medium is by definition shared by all users. The only restrictions which can be imposed are legislative in nature – coming from government regulating agencies such as the Federal Communications Commission (FCC) in the United States and its counterparts in other countries. Specifically, the oldest method of interference mitigation in wireless communication systems is licensing with implied exclusive use of licensed spectrum. However, high licensing fees, along with the need for good returns on large investments, compel wireless service providers to serve as many customers as possible in their licensed band – which even then leads inexorably to mutually interfering users. Thus, mutual interference is a fact of wireless life.

Traditional approaches to combating interference usually begin with measurement and/or prediction of the channel followed by appropriate selection of modulation methods and signal processing algorithms for reliable reception – each carrying an associated set of standards and a relatively rigid hardware infrastructure. That is, historically, once a scheme has been selected, changing to a different one was not an easy option owing to the relatively high complexity of the transmitter and receiver hardware. For example, a system using quadrature amplitude modulation (QAM) cannot be simply changed to a phase shift keying (PSK) system.

However, the emergence of software radios[1] [1, 36, 61, 62] is changing the way modern communication systems are designed by providing programmable radios which can produce different output waveforms and which can act as different receiver types. While radios of this type are not yet widely available, one can now start to imagine instructing transmitting and receiving radios to adapt their modulation schemes to better suit the operating environment. This versatility of transmitter and receiver structures implied by software radio architectures will profoundly affect the methods by which future wireless communication systems deal with interference.

In this book we assume the existence of appropriate software radios, and investigate distributed adaptive methods to maximize performance where each user (subject to power constraints, other-user interference and background noise) will greedily seek to maximize its own signal-to-interference plus noise-ratio (SINR) based on feedback about interference conditions at the receiver. Surprisingly enough, we will find that rather than leading to chaos and poor performance, such greedy distributed methods often lead to optimized use of the shared medium. In fact, under simple assumptions about mutual interference in a synchronous CDMA cellular system, one could expect increases by factors of two to five in the number of supportable users [55]. Moreover, Interference Avoidance (IA) methods, when couched in a general signal space formulation, can be applied to a wide variety of communications settings.

1.   Binary Detection, Karhunen-Loève, and Signal Space

The concept of interference avoidance arises naturally from classical work on whitening filters [72]. Connections with modern methods such as multiuser detection [76] are also readily apparent. Here we provide the necessary background material and note that 1) a similar development may be found in [55] and 2) the signal-space savvy reader can skip this section and proceed with Section 2.

In any case, let us consider the basic model of a digital communication system presented in Figure 1.1, in which a given user transmits digital information in binary form over a noisy channel by transmitting the signal $b\sqrt{p}S(t)$ during a time slot of duration T, where $S(t)$ represents the unit energy signature waveform associated with the given user, p is the received power of the user, and b is the input symbol which can take on values $b = \pm 1$ with equal probability. The time interval $0 \leq t \leq T$ is called the symbol or signaling interval. We assume that the channel corrupts the signal by the addition of an independent stochastic interference waveform $Z(t)$, which is composed in general of both noise that occurs naturally (such as thermal noise), and interfering signals from other

---

[1] Also known as universal radios.

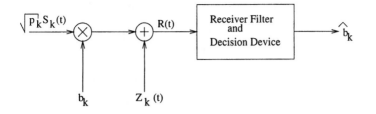

Figure 1.1. Binary digital communication system with one user.

users having different signature waveforms. Therefore, the receiver recovers the signal

$$R(t) = b\sqrt{p}S(t) + Z(t) \tag{1.1}$$

and must decide what symbol was sent by the user. We note that when signature waveforms associated with all users in the system are known, then multiuser detection techniques [76] can be used to design receivers that minimize the probability of error.

In order to derive an estimate $\hat{b}$, most decision devices work with a set of uncorrelated (and preferably independent) variables, which represent sufficient statistics for the received signal $R(t)$, and which can be combined to produce a minimum probability of error estimate for b. Since from the perspective of the receiver $Z(t)$ is a regular stochastic process these variables are obtained through an orthonormal expansion of the interfering waveform $Z(t)$ following the general methodology presented in detail in [72, Ch. 3].

The procedure is based on approximating $Z(t)$ by a finite superposition of orthonormal deterministic functions $\Phi_n(t)$ scaled by uncorrelated random variables a_n

$$\hat{Z}_N(t) = \sum_{n=1}^{N} a_n \Phi_n(t) \tag{1.2}$$

with

$$a_n = <Z(t), \Phi_n(t)> = \int_0^T Z(t)\Phi_n(t)dt \tag{1.3}$$

We define the approximation $\hat{Z}_N(t)$ as good on $[0, T]$ if

$$\text{l.i.m}_{N\to\infty}\hat{Z}_N(t) = Z(t) \tag{1.4}$$

which means that $\hat{Z}_N(t)$ converges to $Z(t)$ in the mean square sense

$$\lim_{N\to\infty} E[(\hat{Z}_N(t) - Z(t))^2] = 0 \tag{1.5}$$

Since the coefficients a_n in equation (1.2) are uncorrelated we have

$$
\begin{aligned}
E[a_n a_k] &= E[< Z(t), \Phi_n(t) >< Z(t), \Phi_k(t) >] \\
&= E\left[\int_0^T \int_0^T Z(t)\Phi_n(t)Z(\tau)\Phi_k(\tau)dtd\tau \right] = \lambda_n \delta_{nk}
\end{aligned}
\tag{1.6}
$$

which is equivalent to

$$
\int_0^T \Phi_n(t) \int_0^T R_Z(t,\tau)\Phi_k(\tau)dtd\tau = \lambda_n \delta_{nk}
\tag{1.7}
$$

where $R_Z(t,\tau) = E[Z(t)Z(\tau)]$ represents the autocorrelation function of the stochastic process $Z(t)$. The relation defined in (1.7) is true only if functions $\Phi_n(t)$ satisfy the Karhunen-Loève integral equation with $R_Z(t,\tau)$ as kernel

$$
\int_0^T R_Z(t,\tau)\Phi_n(\tau)d\tau = \lambda_n \Phi_n(t)
\tag{1.8}
$$

which implies that $\Phi_n(t)$ are eigenfunctions of the stochastic process $Z(t)$ and λ_n are the corresponding eigenvalues. We note that for $n > 2WT$, where W is the communication bandwidth, the eigenvalues λ_n in equation (1.8) rapidly approach zero [72, p. 193]. Thus, the *signal space* of interest implied by finite symbol duration T and finite communication bandwidth W has approximately finite dimension $N \simeq 2WT$ [31].

Now, consider some arbitrary basis set $\{\Psi_i(t)\}$ with which we represent our signal and which also spans the signal space of $Z(t)$. We can define

$$
\Phi_n(t) = \sum_{i=1}^N \varphi_{ni} \Psi_i(t)
\tag{1.9}
$$

with coefficients calculated from

$$
\varphi_{ni} = \int_0^T \Phi_n(t)\Psi_i(t)dt
\tag{1.10}
$$

This representation can be used to develop a discrete equivalent of the Karhunen-Loève integral equation (1.8) as was done in [55], which is helpful since it allows the use of simple linear algebraic methods. By combining equations (1.8) and (1.9) we get

$$
\lambda_n \sum_{i=1}^N \varphi_{ni} \Psi_i(t) = \int_0^T R_Z(t,\tau) \sum_{i=1}^N \varphi_{ni} \Psi_i(\tau)d\tau
\tag{1.11}
$$

and projecting both sides of equation (1.11) onto one of the basis functions $\Psi_k(t)$ yields

$$\lambda_n \varphi_{nk} = \sum_{n=1}^{N} \varphi_{ni} \underbrace{\int_0^T \int_0^T R_Z(t, \tau) \Psi_k(t) \Psi_i(\tau) dt d\tau}_{r_{ki}} \qquad (1.12)$$

The double integral that appears in (1.12) is a quantity that can be calculated based on the autocorrelation function of the stochastic process. Note that r_{ki} can also be rewritten in terms of the projections of Z(t) onto $\{\Psi_n(t)\}$, that is $z_n = \int_0^T Z(t) \Psi_n(t) dt$, as

$$
\begin{aligned}
r_{kn} &= \int_0^T \int_0^T R_Z(t, \tau) \Psi_k(t) \Psi_n(\tau) dt d\tau \\
&= E \left[\int_0^T \int_0^T Z(t) Z(\tau) \Psi_k(t) \Psi_n(\tau) dt d\tau \right] = E[z_k z_n]
\end{aligned}
\qquad (1.13)
$$

Therefore equation (1.12) can be written as

$$\lambda_n \varphi_{nk} = \sum_{i=1}^{N} r_{ki} \varphi_{ni} \qquad (1.14)$$

and for $k = 1, \ldots, N$ we obtain the standard matrix eigenvalue equation

$$E\left[\mathbf{z}\mathbf{z}^{\mathsf{T}}\right] \boldsymbol{\phi}_n = \mathbf{R}\boldsymbol{\phi}_n = \lambda_n \boldsymbol{\phi}_n \qquad (1.15)$$

with $\boldsymbol{\phi}_n = [\varphi_{n1} \ldots \varphi_{nN}]^{\mathsf{T}}$ and $\mathbf{z} = [z_1 \ldots z_N]^{\mathsf{T}}$.

It can be easily seen that each eigenvector $\boldsymbol{\phi}_n$ in equation (1.15) corresponds to an eigenfunction in equation (1.8), and the associated eigenvalue λ_n represents the amount of interference energy carried by that eigenfunction. Also, since $R_Z(t, \tau)$ is an autocorrelation function, the matrix $\mathbf{R}$ is symmetric and positive semi-definite, which implies that it has a full set of orthonormal eigenvectors which span $\mathbb{R}^N$.

The receiver takes the signal $R(t)$ in equation (1.1) as input on the interval [0,T], and projects it onto the eigenfunctions $\Phi_n(t)$ obtained by solving the Karhunen-Loève integral equation (1.8) or the equivalent matrix-eigenvector equation (1.15) to obtain the set of uncorrelated random variables $r_n = s_n + z_n$, with $s_n = < bS(t), \Phi_n(t) >$ and $z_n = < Z(t), \Phi_n(t) >$. Since it has been assumed that the $\{\Phi_n(t)\}_{n=1}^N$ span the signal space for $S(t)$, the s_n contain all available information about $bS(t)$. In a multiuser environment the interference process $Z(t)$ is a sum of independent processes, and therefore the marginal distributions for each z_n are well approximated by Gaussian distributions of zero-mean and variance λ_n. Furthermore, the joint distribution of $\{z_n\}_{n=1}^{N}$ is approximated by a multivariate Gaussian distribution which implies that the z_n are also independent since they are by design uncorrelated. Therefore the probability distribution on the received vector $\mathbf{r} = \begin{bmatrix} r_1 & \ldots & r_N \end{bmatrix}^{\top}$ given the vector of signal projections $\mathbf{s} = \begin{bmatrix} s_1 & \ldots & s_N \end{bmatrix}^{\top}$ is

$$p_{\mathbf{r}|\mathbf{s}}(\mathbf{r}|\mathbf{s}) = \prod_{n=1}^{N} \frac{1}{\sqrt{2\pi\lambda_n}} e^{-\frac{|\mathbf{r}-\mathbf{s}|^2}{2\lambda_n}} \tag{1.16}$$

When $S(t)$ is sent for $b = 1$ and $-S(t)$ is sent for $b = -1$ we get the conditional probability distributions

$$p_{\mathbf{r}|1}(\mathbf{r}|1) = \prod_{n=1}^{N} \frac{1}{\sqrt{2\pi\lambda_n}} e^{-\frac{|\mathbf{r}-\mathbf{s}|^2}{2\lambda_n}} \tag{1.17}$$

$$p_{\mathbf{r}|-1}(\mathbf{r}|-1) = \prod_{n=1}^{N} \frac{1}{\sqrt{2\pi\lambda_n}} e^{-\frac{|\mathbf{r}+\mathbf{s}|^2}{2\lambda_n}} \tag{1.18}$$

Since 1 and -1 are equiprobable, the probability of error is minimized when the decision regions on the received vector $\mathbf{r}$ are chosen according to the likelihood ratio test (LRT)

$$\frac{p_{\mathbf{r}|1}(\mathbf{r}|1)}{p_{\mathbf{r}|-1}(\mathbf{r}|-1)} \underset{\substack{< \\ \text{say -1}}}{\overset{\substack{\text{say 1} \\ >}}{\gtrless}} 1 \tag{1.19}$$

After taking the log of both sides and some algebraic manipulation, equation (1.19) is equivalent to the threshold test for optimal detection [72, 39].

$$\sum_{n=1}^{N} \frac{s_n r_n}{\lambda_n} \underset{\substack{< \\ \text{say -1}}}{\overset{\substack{\text{say 1} \\ >}}{\gtrless}} 0 \tag{1.20}$$

and the probability of error is given by

$$p_e = \int_{\sqrt{\beta}/2}^{\infty} \frac{1}{\sqrt{2\pi}} e^{-\frac{x^2}{2}} dx \tag{1.21}$$

with $\beta = \sum_{n=1}^{N} \frac{s_n^2}{\lambda_n}$.

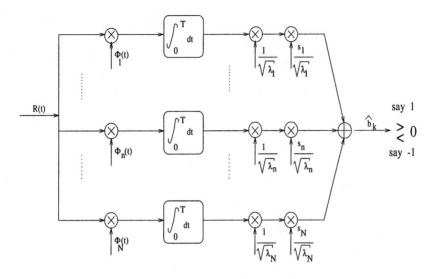

Figure 1.2. Whitening filter receiver

This detection method, whose structure is given in Figure 1.2, is called a *whitening filter* since the decision law in equation (1.20) can be rewritten as

$$\sum_{n=1}^{N} \frac{r_n}{\sqrt{\lambda_n}} \frac{s_n}{\sqrt{\lambda_n}} \underset{\text{say -1}}{\overset{\text{say 1}}{\underset{<}{>}}} 0 \tag{1.22}$$

which represents a rescaling of the received signal vector components yielding interference components with equal energy $z_n/\sqrt{\lambda_n}$ which are similar to a white noise process. A matched filter applied to the rescaled signal vector with components $s_n/\sqrt{\lambda_n}$ completes the detection process. It should be noted that if any of the $\lambda_n = 0$, the formal whitening structure is indeterminate, but optimal detection can still be achieved. Since $\lambda_n = 0$ the n-*th* rail is interference free, and hence if the corresponding $s_n \neq 0$ then perfect detection is possible. On the other hand, if the corresponding $s_n = 0$ then the rail does not carry any information about $bS(t)$ and can be discarded.

2. CDMA and the Interference Avoidance Principle

In a generalized code division multiple access (CDMA) system users transmit signature waveforms $S_k(t)$ that are linear superpositions of the signal space basis functions. In a typical CDMA system, the basis functions might be non-overlapping pulses (or "chips") [23, 48]. In others the basis might be "tones" of different frequencies as in multicarrier and OFDM systems [5], or even spatial signal distributions [2]. In still others one might have a superposition of wavelets [77]. Thus, the representation is completely general in that it can be used with any signal space basis.

We then note that regardless of which basis is used, the detection rule of equation (1.20) corresponds to a receiver filter c with components

$$c_n = \frac{s_n}{\lambda_n} \qquad (1.23)$$

The filter c is immediately recognized as a scaled version of the minimum mean squared error (MMSE) linear filter [33] and the decision law in equation (1.20) represents the MMSE multiuser detector [76, Sec. 6.2]. Thus, the decision variable for $\hat{b}$ is

$$\mathbf{c}^\top \mathbf{r} = \sum_{n=1}^{N} c_n r_n = b \sum_{n=1}^{N} \frac{s_n^2}{\lambda_n} + \sum_{n=1}^{N} \frac{s_n z_n}{\lambda_n} \qquad (1.24)$$

and contains both signal and interference terms. Thus, we can compute the SINR at the output of the MMSE filter as

$$\mathrm{SINR} = \frac{E\left[\left(b \sum_{n=1}^{N} \frac{s_n^2}{\lambda_n}\right)^2\right]}{E\left[\left(\sum_{n=1}^{N} \frac{s_n z_n}{\lambda_n}\right)^2\right]} \qquad (1.25)$$

which reduces to [55]

$$\mathrm{SINR} = \sum_{n=1}^{N} \frac{s_n^2}{\lambda_n} \qquad (1.26)$$

a well known result [33]. However, equation (1.26) also suggests the possibility of obtaining a higher value of the output SINR by adjusting the signature components $\{s_n\}$. Specifically, if $S(t)$ is subject to a unit energy constraint which implies $\sum_{n=1}^{N} s_n^2 = 1$, then SINR can be maximized by choosing $s_n = 1$ for any n such that $\lambda_n = \min_k \lambda_k$, in which case we would have $S(t) = \Phi_n(t)$. Alternatively, the signal energy could be distributed in some arbitrary fashion over all those $\Phi_n(t)$ with minimum eigenvalues λ_n. This result has the following simple and intuitive interpretation:

You maximize SINR by placing signal energy where there is least interference.

This procedure has been dubbed *interference avoidance* and its application for a single user with a given interference process is straightforward.

The surprise, however, is that even when an ensemble of users perform independent greedy interference avoidance, the result is almost always optimal resource allocation from a variety of perspectives [55]. The algorithm will be formally introduced in Chapter 2.

3. Performance Metrics

We now introduce a useful metric of mutual interference called the *total squared correlation* and show how it relates to two distinctly different measures of system capacity.

As previously mentioned, the received signal $R(t)$ during any symbol interval is a superposition of individual signature waveforms modulated by corresponding information symbols along with additive background noise. Under the assumption that background noise is white and Gaussian, the simplest way to eliminate interference in a multiuser system is to use orthogonal signaling [76, Ch. 1]. In this case signature waveforms are chosen such that their inner product (cross correlation)

$$< S_i(t), S_j(t) >= \int_0^T S_i(t)S_j(t)dt = \mathbf{s}_i^\top \mathbf{s}_j \qquad (1.27)$$

is equal to zero. Detection proceeds by correlating the received signal with each $\{S_i(t)\}_{i=1}^N$ followed by a threshold rule. Under the assumption of synchrony, orthogonality dictates that the output of the correlator for user k will not be affected by the signal of any other user $i \neq k$ regardless of their relative powers p_i. Thus, probability of error for orthogonal signatures is the same as if transmission occurred over separate channels for each symbol.

3.1 User Capacity

However, it can often occur that the quality of each channel is much higher than needed with an implicit waste of resources. That is, the tolerable SINR might be γ^* and the actual SINR might be $\gamma \gg \gamma^*$. Although a variety of multi-level signal techniques could be employed to further subdivide each of these "channels", it is often more convenient both intellectually and practically to relax signature orthogonality and allow codewords to be correlated to the extent that each can achieve $\gamma \geq \gamma^*$. The maximum number of users which can sustain a given target SINR is called the *user capacity* of the system [79].

We can formalize user capacity in terms of signature vectors $\mathbf{s}_k$ by noting that the interference seen by a given user is the sum of squared correlations with other users. That is the SINR γ_k of user k is

$$\gamma_k = \frac{p_k \mathbf{s}_k^\top \mathbf{s}_k}{\sum_i p_i (\mathbf{s}_i^\top \mathbf{s}_k)^2 - p_k + w_k} \qquad (1.28)$$

where w_k is the energy in the projection of the additive noise onto s_k. If we assume equal power users with unit norm codewords we then have

$$\gamma_k = \frac{1}{\sum_i (s_i^T s_k)^2 - 1 + w_k} \tag{1.29}$$

We find it useful to represent the ensemble of *codewords* $\{s_k\}$ in matrix form as an $N \times L$ matrix S whose columns are the signature sequences

$$S = \begin{bmatrix} | & | & & | \\ s_1 & s_2 & \cdots & s_L \\ | & | & & | \end{bmatrix} \tag{1.30}$$

Thus, the SINR for user k can be rewritten as

$$\gamma_k = \frac{1}{\text{Trace}\left[(SS^T)^2\right] - 1 + w_k} \tag{1.31}$$

which in a white noise background where each w_k has equal power leads to a compact figure of merit for the codeword ensemble called the *total squared correlation*

$$\text{TSC} \equiv \sum_i (s_i^T s_i)^2 = \text{Trace}\left[(SS^T)^2\right] \tag{1.32}$$

Clearly, smaller TSC leads to larger SINR, and it is easy to show that TSC is lower bounded by L^2/N when $L \geq N$ and by L when $L \leq N$. The result was first derived by Welch in [82]. Codeword ensembles which meet these bounds are therefore called *Welch Bound Equality* sequences, or WBE sequences for short.

Now, the user capacity of the system is the maximum number of admissible users at a given common target SINR γ^*. L users are said to be admissible if there exist powers $p_i > 0$ and signature sequences s_i such that each user has an SINR at least as large as γ^*. It has been shown [79] that user capacity is maximized if signature sequences satisfy

$$S^T S = I_L, \quad \text{if } L \leq N \tag{1.33}$$

or

$$SS^T = \frac{L}{N} I_N, \quad \text{if } L > N \tag{1.34}$$

and received powers of the users are are chosen to be the same. Clearly these results agree with the Welch Bounds for equation (1.32) so that a set of sequences which satisfy either (1.33) or (1.34) will also have minimum TSC. Thus in a white noise background, TSC-minimizing sequences are WBE sequences and *vice versa*.

3.2 Sum Capacity

Of course, user capacity is somewhat subjective in that it depends on the performance criterion γ^*. Thus, it is reasonable to examine codeword ensembles in the context of more objective measures like information theoretic capacity. In a white noise background with identical received powers over all users we define the sum capacity as the maximum sum of reliable (error-free) rates over all users given by [73]

$$C_s = \frac{1}{2} \log \left[\det \left(\mathbf{I}_N + \frac{p}{\sigma} \mathbf{S} \mathbf{S}^{\top} \right) \right] = \frac{1}{2} \log \left[\det \left(\mathbf{I}_L + \frac{p}{\sigma} \mathbf{S}^{\top} \mathbf{S} \right) \right] \quad (1.35)$$

and is achieved by an ensemble $\mathbf{S}$ when codewords are chosen such that they meet the Welch Bound with equality [56].

Thus, not only is TSC a measure of user capacity, but in many cases it can be used as a surrogate for sum capacity. Since TSC is a quadratic function of the codewords, it (and simple variants) will prove a useful analytic and algorithmic tool as we pursue various optimizations in later chapters.

4. Summary

In this chapter we have described the motivation for employing interference avoidance – significant improvements in system capacity and the hope that IA can be an organizing principle for unlicensed band systems in general and in a variety of settings. We described the basic ideas from which interference avoidance springs – whitening filters and linear multiuser detection and have described some of the simple basic metrics we will use to measure performance. We continue in chapter 2 with a careful description of different interference avoidance algorithms and their relationship to square correlation and sum capacity metrics, and in chapter 3 we finish our discussion of interference avoidance on simple channels with the study of algorithm convergence and optimality.

Chapter 2

INTERFERENCE AVOIDANCE ALGORITHMS

In this chapter we will develop the detailed theory behind interference avoidance, consider two important measures of performance and then present a variety of distributed interference avoidance algorithms. We will show that these algorithms converge in the sense that they move the chosen metric toward a finite extremal value. We will defer detailed optimality and convergence discussions and structural properties of optimal codeword ensembles until chapter 3.

To begin, let us consider the uplink of a synchronous CDMA communication system with L users having signature waveforms $\{S_\ell(t)\}_{\ell=1}^{L}$ of finite duration T and equal received power at a common receiver (base station). Without loss of generality we assume unit received power for each user. The received signal is

$$R(t) = \sum_{\ell=1}^{L} b_\ell S_\ell(t) + n(t) \tag{2.1}$$

where b_ℓ is the information symbol sent by user ℓ with signature $S_\ell(t)$, and $n(t)$ is an additive Gaussian noise process that corrupts the signal at the receiver. We assume that all signals are representable in an arbitrary N-dimensional signal space, in which each user's signature waveform $S_\ell(t)$ is equivalent to an N-dimensional vector $\mathbf{s}_\ell$ and Gaussian noise process $n(t)$ is equivalent to the noise vector $\mathbf{n}$ with correlation matrix $E[\mathbf{nn}^\top] = \mathbf{W}$. The equivalent received signal vector $\mathbf{r}$ at the base station is then given by the following vector equation

$$\mathbf{r} = \sum_{\ell=1}^{L} b_\ell \mathbf{s}_\ell + \mathbf{n} \tag{2.2}$$

13

As in chapter 1, $\mathbf{S}$ is the $N \times L$ matrix having as columns the user codewords $\mathbf{s}_\ell$

$$\mathbf{S} = \begin{bmatrix} | & & | & & | \\ \mathbf{s}_1 & \cdots & \mathbf{s}_\ell & \cdots & \mathbf{s}_L \\ | & & | & & | \end{bmatrix} \tag{2.3}$$

and the received signal can be rewritten in vector-matrix form as

$$\mathbf{r} = \mathbf{S}\mathbf{b} + \mathbf{n} \tag{2.4}$$

with $\mathbf{b} = [b_1 \ldots b_\ell \ldots b_L]^\mathsf{T}$ containing the symbols sent by users 1 through L. Let us also denote by $\mathbf{R}$ the autocorrelation matrix of the received signal

$$\mathbf{R} = E[\mathbf{r}\mathbf{r}^\mathsf{T}] = \mathbf{S}\mathbf{S}^\mathsf{T} + \mathbf{W} \tag{2.5}$$

A unit norm receiver filter, $\mathbf{c}_k$, is used to estimate the symbol transmitted by a given user k. This estimate is computed as

$$\hat{b}_k = \mathbf{c}_k^\mathsf{T}\mathbf{r} \tag{2.6}$$

so the expression for the signal-to-interference plus noise-ratio (SINR) for user k is

$$\gamma_k = \frac{(\mathbf{c}_k^\mathsf{T}\mathbf{s}_k)^2}{\displaystyle\sum_{\ell=1,\ell\neq k}^{L} (\mathbf{c}_k^\mathsf{T}\mathbf{s}_\ell)^2 + E[(\mathbf{c}_k^\mathsf{T}\mathbf{n})^2]} \tag{2.7}$$

Assuming simple matched filters at the receiver for all users, the SINR for user k in equation (2.7) is expressed as

$$\gamma_k = \frac{(\mathbf{s}_k^\mathsf{T}\mathbf{s}_k)^2}{\displaystyle\sum_{\ell=1,\ell\neq k}^{L} (\mathbf{s}_k^\mathsf{T}\mathbf{s}_\ell)^2 + E[(\mathbf{s}_k^\mathsf{T}\mathbf{n})^2]}$$

$$= \frac{1}{\mathbf{s}_k^\mathsf{T}\left(\displaystyle\sum_{\ell=1,\ell\neq k}^{L} \mathbf{s}_\ell\mathbf{s}_\ell^\mathsf{T} + E[\mathbf{n}\mathbf{n}^\mathsf{T}]\right)\mathbf{s}_k^\mathsf{T}} \tag{2.8}$$

where we have once again assumed that $|s_k| = 1$.[1] Proceeding, we define the autocorrelation matrix of the interference-plus-noise seen by user k

$$\mathbf{R}_k = \sum_{\ell=1,\ell\neq k}^{L} \mathbf{s}_\ell \mathbf{s}_\ell^\top + \mathbf{W} = \mathbf{R} - \mathbf{s}_k \mathbf{s}_k^\top \qquad (2.9)$$

Then, equation (2.8) can be rewritten as

$$\gamma_k = \frac{1}{\mathbf{s}_k^\top \mathbf{R}_k \mathbf{s}_k} \qquad (2.10)$$

The idea behind interference avoidance algorithms is to use the SINR as a metric, and maximize it through adaptation of user codewords. This is also equivalent to minimizing the inverse SINR, defined as

$$\beta_k = \frac{1}{\gamma_k} = \mathbf{s}_k^\top \mathbf{R}_k \mathbf{s}_k \qquad (2.11)$$

Note that for unit norm codewords, equation (2.11) represents the Rayleigh quotient for matrix $\mathbf{R}_k$, and recall from linear algebra [66, p. 348] that this is minimized by the eigenvector corresponding to the minimum eigenvalue of the given matrix. Thus, the SINR for user k can be greedily maximized by having user k replace its current codeword $\mathbf{s}_k$ with the minimum eigenvector of the autocorrelation matrix $\mathbf{R}_k$ of the interference-plus-noise seen by user k. We call this procedure *greedy interference avoidance* since by replacing its current codeword with the minimum eigenvector of the interference-plus-noise correlation matrix, user k avoids interference by placing its transmitted energy in that region of the signal space with minimum interference-plus-noise energy and greedily maximizes SINR without paying attention to potentially negative effects this action may have on other users in the system.

Of course, a pivotal question is whether such an iterative procedure produces optimal codeword ensembles when employed by multiple users. This convergence issue turns out to be a surprisingly difficult one to address rigorously. Nonetheless, we will start by revisiting the TSC and sum capacity metrics introduced in chapter 1 and modifying them to include general background noise covariance $\mathbf{W}$.

Specifically, we define *general squared correlation* (GSC) as

$$\text{GSC} = \text{Trace}\left[\mathbf{R}^2\right] = \text{Trace}\left[\left(\mathbf{S}\mathbf{S}^\top + \mathbf{W}\right)^2\right] \qquad (2.12)$$

---

[1]For simplicity of exposition we have assumed unit energy codewords. However, it should be noted that all the results we present hold even if unequal codeword energies $|s_k|^2 = p_k$ are assumed [54, 55].

If the noise covariance is white ($\mathbf{W} = \alpha \mathbf{I}$), then

$$
\begin{aligned}
\text{GSC} &= \text{Trace}\left[(\mathbf{S}\mathbf{S}^\top)^2\right] + \alpha^2 \text{Trace}\left[(\mathbf{I})^2\right] + 2\alpha \text{Trace}\left[(\mathbf{S}\mathbf{S}^\top)\right] \\
&= \text{Trace}\left[(\mathbf{S}\mathbf{S}^\top)^2\right] + \alpha^2 N + 2\alpha P
\end{aligned}
\tag{2.13}
$$

where $P = \sum_\ell |s_\ell|^2$ is the total codeword energy. Thus in white noise, TSC is related to GSC by an additive constant. Likewise we define the sum capacity [54, 55] as

$$
C_s = \frac{1}{2} \log(\det \mathbf{R}) - \frac{1}{2} \log(\det \mathbf{W})
\tag{2.14}
$$

We will now determine extremal values for GSC and sum capacity. First, we define the eigenvalues of $\mathbf{R}$ as $\{\lambda_i\}$, $i = 1, \cdots N$, where $\lambda_i > 0$ because $\mathbf{R}$ is positive definite since we assume the noise covariance $\mathbf{W}$ is positive definite. This allows us to rewrite GSC and sum capacity as

$$
\text{GSC} = \sum_i \lambda_i^2
\tag{2.15}
$$

and

$$
C_s = \frac{1}{2} \sum_i \log \lambda_i - \frac{1}{2} \log(\det \mathbf{W})
\tag{2.16}
$$

respectively. To determine extremal values we note the constraint $\text{Trace}\,[\mathbf{R}] = \sum_i \lambda_i = P + \text{Trace}\,[\mathbf{W}]$, a constant. Performing the constrained optimization for GSC with Lagrange multipier κ_1 yields

$$
\frac{\partial}{\partial \lambda_i} \left[\text{GSC} + \kappa_1 \left(\text{Trace}\,[\mathbf{R}] - \sum_i \lambda_i \right) \right] = 0
\tag{2.17}
$$

so that

$$
\lambda_i = \frac{\kappa_1}{2}
\tag{2.18}
$$

Since

$$
\frac{\partial^2}{\partial \lambda_i \partial \lambda_j} \left[\text{GSC} + \kappa_1 \left(\text{Trace}\,[\mathbf{R}] - \sum_i \lambda_i \right) \right] = 2\delta_{ij}
\tag{2.19}
$$

the optimization is convex and we know that constant $\lambda_i = \frac{\text{Trace}[\mathbf{W}]+P}{N}$ absolutely minimizes GSC. Likewise for sum capacity we have

$$
\frac{\partial}{\partial \lambda_i} \left[\frac{1}{2} \sum_i \log \lambda_i + \kappa_2 \left(\text{Trace}\,[\mathbf{R}] - \sum_i \lambda_i \right) \right] = 0
\tag{2.20}
$$

so that

$$\lambda_i = \frac{2}{\kappa_2} \tag{2.21}$$

and similar to GSC we have

$$\frac{\partial^2}{\partial \lambda_i \partial \lambda_j} \left[\frac{1}{2} \sum_i \log \lambda_i + \kappa_2 \left(\text{Trace}\,[\mathbf{R}] - \sum_i \lambda_i \right) \right] = \begin{cases} -\dfrac{1}{2} \dfrac{1}{\lambda_i^2} & i = j \\ 0 & i \neq j \end{cases} \tag{2.22}$$

which is again a convex optimization so that the same constant $\{\lambda_i\}$ as for GSC maximize sum capacity.

So, now that we know GSC is bounded from below and sum capacity is bounded from above we may ask the following general question:

Does there exist a class of iterative procedures which moves these metrics toward their respective bounds?

The answer, of course, is yes and since GSC and C_s are bounded, any such procedure is guaranteed to converge. Of course, convergence of a given metric such as GSC or sum capacity does not address the issue of whether codeword *ensembles* actually converge or even whether GSC and sum capacity converge to their extremal values. However, we will defer that issue until chapter 3 and for now simply ask what procedures will nudge TSC and sum capacity in the right direction.

1. Greedy Interference Avoidance: the eigen-algorithm

Consider the following algorithm:

Eigen-Algorithm

1 Start with a randomly chosen codeword ensemble specified by the user codewords $\{\mathbf{s}_\ell\}_{\ell=1}^L$

2 For $k = 1, \ldots, L$

    (a) Compute the autocorrelation matrix $\mathbf{R}_k$ of the interference-plus-noise seen by user k

    (b) Determine the minimum eigenvalue $\lambda_N^{(k)}$ of $\mathbf{R}_k$ and its associated unit eigenvector $\mathbf{v}_N^{(k)}$.

    (c) If user k's codeword $\mathbf{s}_k$ is not already a suitable eigenvector of $\mathbf{R}_k$, replace it by $\mathbf{v}_N^{(k)}$.

3 Repeat step 2

Since each codeword is iteratively replaced by a minimum eigenvector, this algorithm was dubbed *the eigen-algorithm* in [55].

We will first show that the eigen-algorithm monotonically decreases GSC. From equation (2.9) we have

$$\mathbf{R} = \mathbf{R}_k + \mathbf{s}_k \mathbf{s}_k^\top \qquad (2.23)$$

and when user k replaces its codeword $\mathbf{s}_k$ with a new codeword $\mathbf{x}$ the difference in GSC can be written as

$$\Delta = \text{Trace}\left[(\mathbf{R}_k + \mathbf{s}_k \mathbf{s}_k^\top)^2\right] - \text{Trace}\left[(\mathbf{R}_k + \mathbf{x}_k \mathbf{x}_k^\top)^2\right] \qquad (2.24)$$

This can be further rewritten after canceling similar terms and replacing the traces by the corresponding quadratic forms as

$$\Delta = 2(\mathbf{s}_k^\top \mathbf{R}_k \mathbf{s}_k - \mathbf{x}_k^\top \mathbf{R}_k \mathbf{x}_k) \qquad (2.25)$$

and when $\mathbf{x}$ is chosen to be the minimum eigenvector of $\mathbf{R}_k$ we have that

$$\mathbf{s}_k^\top \mathbf{R}_k \mathbf{s}_k \geq \mathbf{x}_k^\top \mathbf{R}_k \mathbf{x}_k \qquad (2.26)$$

which implies that $\Delta \geq 0$. Thus, the eigen-algorithm monotonically decreases GSC.

Now consider that sum capacity in equation (2.14) is increased if

$$\frac{1}{2} \log \det \left(\mathbf{R}_k + \mathbf{x}_k \mathbf{x}_k^\top\right) \geq \frac{1}{2} \log \det \left(\mathbf{R}_k + \mathbf{s}_k \mathbf{s}_k^\top\right) \qquad (2.27)$$

Since $\mathbf{R}_k$ is assumed invertible, we can factor out $\det \mathbf{R}_k$ in equation (2.27) to obtain

$$\det \left(\mathbf{I} + \mathbf{R}_k^{-\frac{1}{2}} \mathbf{x}_k \mathbf{x}_k^\top \mathbf{R}_k^{-\frac{1}{2}}\right) \geq \det \left(\mathbf{I} + \mathbf{R}_k^{-\frac{1}{2}} \mathbf{s}_k \mathbf{s}_k^\top \mathbf{R}_k^{-\frac{1}{2}}\right) \qquad (2.28)$$

Since the matrices $\mathbf{R}_k^{-\frac{1}{2}} \mathbf{s}_k \mathbf{s}_k^\top \mathbf{R}_k^{-\frac{1}{2}}$ and $\mathbf{R}_k^{-\frac{1}{2}} \mathbf{x}_k \mathbf{x}_k^\top \mathbf{R}_k^{-\frac{1}{2}}$ each have rank one, equation (2.28) further reduces to

$$(1 + \mathbf{x}_k^\top \mathbf{R}_k^{-1} \mathbf{x}_k) \geq (1 + \mathbf{s}_k^\top \mathbf{R}_k^{-1} \mathbf{s}_k) \qquad (2.29)$$

or

$$\mathbf{x}_k^\top \mathbf{R}_k^{-1} \mathbf{x}_k \geq \mathbf{s}_k^\top \mathbf{R}_k^{-1} \mathbf{s}_k \qquad (2.30)$$

which is strikingly similar to, *but different from*, equation (2.26). Of particular note, equations (2.26) and (2.30) taken together imply that one could devise an algorithm which though it always improved one metric, might sometimes degrade the other. More precisely, we emphasize that equations (2.26) and (2.30) are *not equivalent inequalities*. Nonetheless, the same logic applies in regard to the eigen-algorithm. If $\mathbf{x}_k$ is chosen as the minimum eigenvector of $\mathbf{R}_k$, then $\mathbf{x}_k$ is also the maximum eigenvector of $\mathbf{R}_k^{-1}$. Thus, via the Rayleigh quotient and equation (2.30), the eigen-algorithm cannot decrease sum capacity.

So in summary, the eigen-algorithm decreases GSC and increases sum capacity at each step. Since GSC and sum capacity are bounded from below and above respectively, the eigen-algorithm must converge to some value. Once again, we point out that such convergence does not imply convergence of codewords or even convergence to extremal values of GSC and sum capacity. We have only shown that the iterative procedure must converge in GSC and sum capacity.

2. MMSE Interference Avoidance

An alternative interference avoidance procedure can be defined using the MMSE (minimum mean square error) receiver filter as the codeword replacement vector.[2] The MMSE filter is obtained [33] by minimizing the mean squared error (MSE) between the filter output and the transmitted information symbol

$$\text{MSE}_k = E[(\mathbf{c}_k^\top \mathbf{r} - b_k)^2] = \mathbf{c}_k^\top \mathbf{R}_k \mathbf{c}_k + (\mathbf{c}_k^\top \mathbf{s}_k - 1)^2 \tag{2.31}$$

The MMSE filter for user k is then

$$\mathbf{c}_k = \arg \min_{\mathbf{c}_k} \text{MSE}_k \tag{2.32}$$

and since MSE_k is a quadratic function in $\mathbf{c}_k$ the necessary and sufficient condition for optimal $\mathbf{c}_k$ is

$$\frac{\partial}{\partial \mathbf{c}_k}(\text{MSE}_k) = 0 \tag{2.33}$$

which yields

$$2\mathbf{R}_k \mathbf{c}_k + 2\mathbf{s}_k(\mathbf{s}_k^\top c_k - 1) = 0 \tag{2.34}$$

from where

$$\mathbf{c}_k = (\mathbf{R}_k + \mathbf{s}_k \mathbf{s}_k^\top)^{-1} \mathbf{s}_k = \mathbf{R}^{-1} \mathbf{s}_k \tag{2.35}$$

Using the matrix inversion lemma [27, p. 19] equation (2.35) can be rewritten as

$$\mathbf{c}_k = \frac{\mathbf{R}_k^{-1} \mathbf{s}_k}{1 + \mathbf{s}_k^\top \mathbf{R}_k^{-1} \mathbf{s}_k} \tag{2.36}$$

We note that $\mathbf{c}_k$ in equation (2.35) does not have unit norm, and with the appropriate normalization we obtain the expression of the unit norm MMSE receiver filter for user k

$$\mathbf{c}_k = \frac{\mathbf{R}_k^{-1} \mathbf{s}_k}{(\mathbf{s}_k^\top \mathbf{R}_k^{-2} \mathbf{s}_k)^{1/2}} \tag{2.37}$$

---

[2]Historically, MMSE interference avoidance was discovered first [71].

We also note that the MMSE receiver filter maximizes the SINR [33] in equation (2.7). Thus, when MMSE receiver filters are assumed, the SINR for user k can be maximized by having user k replace its codeword with the normalized MMSE receiver filter corresponding to s_k. Therefore, this method was dubbed the *MMSE update* in [71] and defines an alternative interference avoidance procedure. The formal MMSE codeword replacement algorithm is then:

The MMSE Algorithm

1 Start with a randomly chosen codeword ensemble specified by the user codewords $\{s_\ell\}_{\ell=1}^L$.

2 For $k = 1, \ldots, L$

    (a) Compute the normalized MMSE receiver filter c_k for user k using equation (2.37)

    (b) Replace user k codeword s_k by c_k

3 Repeat step 2

The fact that this interference avoidance procedure based on the MMSE codeword update monotonically decreases TSC and increases sum capacity in a white noise background was proven in [71]. However, for GSC and a general noise background the proof is slightly more complex since we need to show that for the MMSE update we have

$$s_k^T R_k s_k \geq c_k^T R_k c_k \qquad (2.38)$$

which is equivalent to

$$s_k^T R_k s_k \geq \left(\frac{R_k^{-1} s_k}{(s_k^T R_k^{-2} s_k)^{1/2}} \right)^T R_k \frac{R_k^{-1} s_k}{(s_k^T R_k^{-2} s_k)^{1/2}} \qquad (2.39)$$

or

$$s_k^T R_k^{-1} s_k \leq \left(s_k^T R_k^{-2} s_k \right) \left(s_k^T R_k s_k \right) \qquad (2.40)$$

Since R_k is the covariance matrix of the interference plus noise seen by user k it is symmetric and positive definite. Therefore it is invertible and since s_k is unit norm we can write

$$\|s_k\|^2 = s_k^T R_k^{-1/2} R_k^{1/2} s_k = 1 \qquad (2.41)$$

Applying the Schwarz inequality [66, p. 147] we have

$$1 = \left(s_k^T R_k^{-1/2} R_k^{1/2} s_k \right)^2 \leq \|R_k^{-1/2} s_k\|^2 \|R_k^{1/2} s_k\|^2 \qquad (2.42)$$

that is

$$1 \le \left(\mathbf{s}_k^\top \mathbf{R}_k^{-1} \mathbf{s}_k \right) \left(\mathbf{s}_k^\top \mathbf{R}_k \mathbf{s}_k \right) \tag{2.43}$$

Furthermore, using the Schwarz inequality again we have

$$\left(\mathbf{s}_k^\top \mathbf{R}_k^{-1} \mathbf{s}_k \right)^2 \le \|\mathbf{s}_k\|^2 \|\mathbf{R}_k^{-1} \mathbf{s}_k\|^2 = \mathbf{s}_k^\top \mathbf{R}_k^{-2} \mathbf{s}_k \tag{2.44}$$

Using equations (2.43) and (2.44) we get

$$\mathbf{s}_k^\top \mathbf{R}_k^{-1} \mathbf{s}_k \le \left(\mathbf{s}_k^\top \mathbf{R}_k^{-1} \mathbf{s}_k \right)^2 \left(\mathbf{s}_k^\top \mathbf{R}_k \mathbf{s}_k \right) \le \left(\mathbf{s}_k^\top \mathbf{R}_k^{-2} \mathbf{s}_k \right) \left(\mathbf{s}_k^\top \mathbf{R}_k \mathbf{s}_k \right) \tag{2.45}$$

which proves that GSC is monotonically decreased for the MMSE codeword update as well.

To show the MMSE algorithm increases sum capacity we return to equation (2.30) and replace $\mathbf{x}$ by the expression for $\mathbf{c}_k$ in equation (2.37) and pose the question

$$\frac{\mathbf{s}_k^\top \mathbf{R}_k^{-3} \mathbf{s}_k}{\mathbf{s}_k^\top \mathbf{R}_k^{-2} \mathbf{s}_k} \overset{?}{\ge} \mathbf{s}_k^\top \mathbf{R}_k^{-1} \mathbf{s}_k \tag{2.46}$$

We then rewrite $\mathbf{s}_k$ as $\Phi \mathbf{z}_k$ where

$$\mathbf{R}_k^{-1} = \Phi \Lambda^{-1} \Phi^\top$$

Our question then becomes

$$\mathbf{z}_k^\top \Lambda^{-1} \mathbf{z}_k \overset{?}{\le} \mathbf{z}_k^\top \Lambda^{-3} \mathbf{z}_k / \mathbf{z}_k^\top \Lambda^{-2} \mathbf{z}_k$$

Rearranging we have

$$\mathbf{z}_k^\top \Lambda^{-3} \mathbf{z}_k - \mathbf{z}_k^\top \Lambda^{-2} \mathbf{z}_k \mathbf{z}_k^\top \Lambda^{-1} \mathbf{z}_k \overset{?}{\ge} 0$$

and factoring yields

$$\mathbf{z}_k^\top \left[\Lambda^{-2} \left(\mathbf{I} - \mathbf{z}_k \mathbf{z}_k^\top \right) \Lambda^{-1} \right] \mathbf{z}_k \overset{?}{\ge} 0$$

Since $\mathbf{I} - \mathbf{z}_k \mathbf{z}_k^\top$ is positive semi-definite and Λ is positive definite, the expression must be greater than or equal to zero. So in general, the MMSE algorithm does not decrease sum capacity.

Thus, sequential application of the MMSE codeword update by all users in the system defines the MMSE algorithm for interference avoidance [55, 71]. As with the eigen-algorithm, convergence of the MMSE algorithm is guaranteed by the monotonic decrease of GSC and increase of sum capacity coupled to the lower and upper bounds, respectively, of GSC and C_s. And once again we emphasize that convergence in these metrics *does not* guarantee either codeword ensemble convergence or even convergence to extremal values of the metrics.

3. Variations On A Theme

We have presented two different interference avoidance algorithms, both of which increase sum capacity and decrease GSC at each step. However, we have also identified a general class of algorithm with guaranteed convergence. That is, any replacement procedure for which the new codeword $\mathbf{x}_k$ satisfies either

$$\mathbf{x}_k^\top \mathbf{R}_k \mathbf{x}_k \leq \mathbf{s}_k^\top \mathbf{R}_k \mathbf{s}_k \tag{2.47}$$

or

$$\mathbf{x}_k^\top \mathbf{R}_k^{-1} \mathbf{x}_k \geq \mathbf{s}_k^\top \mathbf{R}_k^{-1} \mathbf{s}_k \tag{2.48}$$

will result in a convergent algorithm when iterated over all users.

One obvious choice which will later have practical applications is to incrementally adjust the current codeword in either the direction of the optimal codeword, or in such a way that C_s is increased or GSC is decreased. The following sections introduce two such procedures: *lagged IA* and *gradient descent IA*.

3.1 Lagged IA

Starting from the eigen-algorithm, the most obvious of these incremental procedures can be called "lagged IA" with codeword update equation

$$\mathbf{s}_k(t+1) = \frac{\alpha \mathbf{s}_k(t) + m\beta \mathbf{s}_k^*}{|\alpha \mathbf{s}_k(t) + m\beta \mathbf{s}_k^*|} \tag{2.49}$$

where $\alpha, \beta \in \mathbb{R}^+$ and $\mathbf{s}_k^*$ is the minimum eigenvector of $\mathbf{R}_k$, the interference covariance seen by user k at time step t. To ensure that codewords change incrementally we require $|\alpha| \gg |\beta|$, and we explicitly include a time index t to emphasize the incremental nature of the process. We have defined $m = \text{sgn}[\rho_k(t)]$ with $\rho_k(t) = \mathbf{s}_k^\top(t)\mathbf{s}_k^*$ and note that $|\rho_k(t)| \leq 1$ since $|\mathbf{s}_k^*| = |\mathbf{s}_k| = 1$. Equation (2.49) has a simple and intuitive geometric meaning: $\mathbf{s}_k(t+1)$ represents a step toward the closest optimal codeword $m\mathbf{s}_k^*$ along the arc joining $m\mathbf{s}_k^*$ and $\mathbf{s}_k$. That is, $\mathbf{s}_k(t)$ and $m\mathbf{s}_k^*$ share the same half-space.

We now show that GSC is decreased by the iteration as defined in equation (2.49). To this end, let $[(\lambda_1, \phi_1), (\lambda_2, \phi_2), .., (\lambda_L, \phi_L)]$ be the eigenvalues and eigenvectors of $\mathbf{R}_k$ with the λ_i ordered as $\lambda_1 \leq \lambda_2 \leq ... \leq \lambda_L$ ($\mathbf{s}_k^* = \phi_1$). Thus

$$\mathbf{s}_k^\top(t)\mathbf{R}_k \mathbf{s}_k(t) \geq (\mathbf{s}_k^*)^\top \mathbf{R}_k \mathbf{s}_k^* = \lambda_1 \tag{2.50}$$

since $|\mathbf{s}_k(t)| = |\mathbf{s}_k^*| = 1$. The change in GSC is

$$
\begin{aligned}
\Delta\chi_k &= \mathbf{s}_k^\top(t+1)\mathbf{R}_k\mathbf{s}_k(t+1) - \mathbf{s}_k^\top(t)\mathbf{R}_k\mathbf{s}_k(t) \\
&= \frac{(\alpha\mathbf{s}_k(t) + m\beta\mathbf{s}_k^*)^\top}{|\alpha\mathbf{s}_k(t) + m\beta\mathbf{s}_k^*|}\mathbf{R}_k\frac{(\alpha\mathbf{s}_k(t) + m\beta\mathbf{s}_k^*)}{|\alpha\mathbf{s}_k(t) + m\beta\mathbf{s}_k^*|} - \mathbf{s}_k(t)^\top\mathbf{R}_k\mathbf{s}_k(t)
\end{aligned}
$$
(2.51)

We then note that

$$
|\alpha\mathbf{s}_k(t) + m\beta\mathbf{s}_k^*|^2 = \alpha^2 + \beta^2 + 2\alpha\beta|\rho(t)| \equiv \kappa^2 \tag{2.52}
$$

so that

$$
\begin{aligned}
\kappa^2\Delta\chi_k &= \alpha^2\mathbf{s}_k^\top(t)\mathbf{R}_k\mathbf{s}_k(t) - \kappa^2\mathbf{s}_k^\top(t)\mathbf{R}_k\mathbf{s}_k(t) \\
&+ m\alpha\beta\mathbf{s}_k^\top(t)\mathbf{R}_k\mathbf{s}_k^* + m\alpha\beta\mathbf{s}_k^{*\top}\mathbf{R}_k\mathbf{s}_k(t) + \beta^2\mathbf{s}_k^{*\top}\mathbf{R}_k\mathbf{s}_k^* \\
&= (\alpha^2 - \kappa^2)\mathbf{s}_k^\top(t)\mathbf{R}_k\mathbf{s}_k(t) + \lambda_1(\kappa^2 - \alpha^2) \\
&= (\kappa^2 - \alpha^2)\left(\lambda_1 - \mathbf{s}_k^\top(t)\mathbf{R}_k\mathbf{s}_k(t)\right)
\end{aligned}
$$
(2.53)

which for $\alpha, \beta > 0$ is always less than or equal to zero by equation (2.52) and equation (2.50). So, the iteration reduces GSC.

An identical derivation with $\mathbf{R}_k^{-1}$ replacing $\mathbf{R}_k$ and λ_i^{-1} replacing λ_i shows that lagged IA also increases sum capacity. So lagged IA converges in both GSC and sum capacity.

3.2 Gradient Descent IA

Another simple method for interference avoidance can be based on gradient descent. Specifically, we'd like to reduce the expression $\mathbf{s}_k^\top\mathbf{R}_k\mathbf{s}_k$ with each iteration. However, we would also like to maintain unit norm codewords while the obvious minimum of $\mathbf{s}_k^\top\mathbf{R}_k\mathbf{s}_k$ is $\mathbf{s}_k = 0$. We could formulate a constrained gradient, but instead we utilize a Rayleigh quotient

$$
\chi_k = \frac{\mathbf{s}_k^\top\mathbf{R}_k\mathbf{s}_k}{\mathbf{s}_k^\top\mathbf{s}_k} \tag{2.54}
$$

which is the inverse SINR for the kth user and take its gradient with respect to the codeword components $\{s_{kj}\}$ to obtain

$$
\nabla\chi_k = \frac{2[\mathbf{s}_k^\top\mathbf{s}_k\mathbf{R}_k\mathbf{s}_k - (\mathbf{s}_k^\top\mathbf{R}_k\mathbf{s}_k)\mathbf{s}_k]}{(\mathbf{s}_k^\top\mathbf{s}_k)^2} \tag{2.55}
$$

Therefore, the iteration $\mathbf{s}_k(t+1) = \mathbf{s}_k(t) - \nu\Delta\chi_k(t)$ with ν a suitably small constant, would increase $SINR$. However, we require unit norm $\mathbf{s}_k$ so we form

$$\mathbf{s}_k(t+1) = \frac{\mathbf{s}_k(t) - \nu\nabla\chi_k(t)}{\|\mathbf{s}_k(t) - \nu\nabla\chi_k(t)\|} \tag{2.56}$$

This iteration decreases $\mathbf{s}_k^\top \mathbf{R}_k \mathbf{s}_k$ while maintaining a unit norm $\mathbf{s}_k$. Thus, it decreases GSC. That is

$$\mathbf{s}_k^\top(t+1)\mathbf{R}_k\mathbf{s}_k(t+1) \leq \mathbf{s}_k^\top(t)\mathbf{R}_k\mathbf{s}_k(t) \tag{2.57}$$

Therefore convergence is guaranteed. We note that gradient descent interference avoidance does not explicitly require calculation of the minimum eigenvector which is a distinct computational advantage. However, as with any incremental method, convergence will proceed more slowly than if the optimal codeword where chosen straight away at each step [14].

4.   Summary

We have presented four different distributed interference avoidance algorithms, all of which are guaranteed to converge in GSC and/or sum capacity. We have shown that absolutely maximizing sum capacity also absolutely minimizes GSC if the eigenvalues of the covariance $\mathbf{R}$ are only constrained by their sum – which hints that sum capacity maximization and GSC minimization might be equivalent optimizations. However, we have not yet shown codeword convergence and more importantly, we have not shown convergence to optimal values of GSC and sum capacity. In fact, we have not even *determined* the optimal values of these quantities in general, but are in effect simply taking on faith that greedily improving performance, one codeword at a time, is a useful thing to do. As one might expect, it turns out that such greedy and uncoordinated but asynchronous behavior is indeed a good thing to do as we shall see as we carefully explore these issues in chapter 3.

Chapter 3

FIXED POINTS, CONVERGENCE
AND OPTIMALITY

In chapter 2 we found that both GSC and sum capacity have finite extremal values and that there exist a variety of iterative greedy interference avoidance (IA) algorithms which decrease GSC and/or increase sum capacity. Thus, IA must converge in GSC and/or sum capacity. But we left open the key issues of codeword ensemble convergence and convergence to extremal values of the chosen metric.

In this chapter we explicitly consider optimality and codeword convergence. We will find that the extremal values associated with GSC and sum capacity is often a waterfilling solution [4, 15, 54, 55, 78, 79] and that minimizing GSC is *completely* equivalent to maximizing sum capacity. We will carefully describe the characteristics of optimal solutions and then finally will show that codeword ensembles must converge, but in a special manner that we will call *convergence in class*. We will close with numerical examples of interference avoidance and pictorial examples to illustrate various structural properties of optimal codeword ensembles.

1.   Optimal Covariance Structure

We saw in chapter 2 that GSC and sum capacity are minimized and maximized, respectively, if all the eigenvalues of $\mathbf{R}$ are equal. Thus, absolutely minimizing GSC is equivalent to absolutely maximizing C_s. However, the result only holds for cases where the eigenvalues can be made equal via suitable choice of codeword matrix $\mathbf{S}$. We therefore now review more rigorous conditions on optimality for GSC and sum capacity and show that these two optimizations are indeed equivalent without restriction.

First we define the eigenvalues of the noise covariance $\mathbf{W}$ as $\{\sigma_i\}$, $i = 1, \ldots, N$ and assume they are ordered from largest to smallest. Likewise we now allow for differing values of codeword norms $p_k = |\mathbf{s}_k|^2$ and also assume these L $\{p_k\}$ ($L \geq N$) are ordered from largest to smallest. We will always assume $L \geq N$ since if otherwise, the codewords will naturally reside in the $L < N$ noise subspace with smallest σ_i [54, 55]. We note that the interference avoidance procedures and metrics we have already derived do not change so long as each codeword is replaced by a new codeword with the same norm [54, 55]. We then define the eigenvalues of the covariance matrix $\mathbf{R}$ as $\{\lambda_i\}$ this time ordered from largest to smallest. We then define total codeword power as

$$P = \sum_{k=1}^{L} p_k \tag{3.1}$$

and total noise power as

$$U = \sum_{n=1}^{N} \sigma_n \tag{3.2}$$

Since both GSC and sum capacity depend only on the eigenvalues of the covariance matrix $\mathbf{R}$, we must find bounds on the eigenvalues $\{\lambda_i\}$. In fact, we will find bounds on partial sums of these eigenvalues starting with λ_1 and then show that GSC is minimized and sum capacity maximized when this partial sum bound is met. To begin we note, somewhat obviously, that $\lambda_1 \geq \sigma_1$. Equally obvious we note that $(P + U)/N$ is the minmax value of λ_1 since it corresponds to the all-equal eigenvalue solution and λ_1 is assumed the largest of all eigenvalues. Slightly less obvious is the fact that

$$\lambda_1 \geq \frac{1}{k} \sum_{i=1}^{k} (p_i + \sigma_{N-i+1}) \tag{3.3}$$

for $k = 1, 2, \ldots, N - 1$. but becomes obvious when we note that this is the minmax value of the eigenvalues associated with the k largest codeword powers packed into the k least noisy dimensions. All told, we then have the following bound on λ_1 [54]

$$\lambda_1 \geq \max \left[\sigma_1, \max_{0<k<N} \frac{1}{k} \sum_{i=1}^{k} (p_i + \sigma_{N-i+1}), \frac{P+U}{N} \right] \tag{3.4}$$

which, as explained, contains a wealth of structural information in a compact notation.

Furthermore, the bound of equation (3.4) can be used recursively to generate a lower bound on partial sums of eigenvalues $\sum_{i=1}^{n} \lambda_i$ [54]. Examples will help illustrate the procedure. Consider the simplest case where $p_k = 1$ and the noise is white so that $\sigma_i = \sigma$. By applying equation (3.4) we have $\lambda_1 \geq \frac{P+U}{N}$. If we set λ_1 equal to the bound, then all the other λ_i must be equal to $\frac{P+U}{N}$

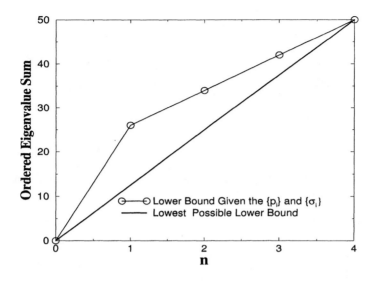

Figure 3.1. Illustration of $\sum_{i=1}^{n} \lambda_i$ bounds in n for text example.

since the sum of eigenvalues is Trace $[\mathbf{R}] = P + U$ and λ_1 is assumed a largest eigenvalue. We note that this equal-eigenvalue constellation corresponds to the absolute minimum GSC and absolute maximum sum capacity following the results of our Lagrange optimizations equation (2.17) and equation (2.20).

As another example consider, $\underline{\sigma} = (26, 6, 4, 4)$ and $\underline{p} = (4, 3, 2, 1)$. Application of equation (3.4) has $\lambda_1 \geq \sigma_1 = 26$. We set λ_1 equal to the bound and then continue with a reduced set of noise eigenvalues in the three remaining dimensions which leaves us with $\underline{\sigma}' = (6, 4, 4)$ and $\lambda_1 + \lambda_2 \geq 34 = 26 + p_1 + \sigma_3'$. Again setting λ_2 equal to the bound indicated and therefore removing p_1 and σ_3' from consideration, we are left with $\underline{\sigma}'' = (6, 4)$ and $\underline{p}'' = (3, 2, 1)$. Applying the bound again we have $\lambda_3 + \lambda_2 + \lambda_1 \geq 42 = 34 + (p_1 + p_3'' + p_3'' + \sigma_1'' + \sigma_2'')/2$. As yet another example[1], consider $\underline{\sigma} = (1.1, 1, 0)$ and $\underline{p} = (1, 1, 1/5)$. In this case $\lambda_1 \geq 3/2$ and $\lambda_1 + \lambda_2 \geq 3$.

In general, it is clear that when the bound for $\sum_{i=1}^{n} \lambda_i$ is plotted against n, the result must be a concave segmented arc above the line $n(P + U)/N$ as illustrated in FIGURE 3.1 for the first example. The absolute (but possibly unattainable) lower bound corresponds to the equal-eigenvalue partial sum plot.

The key result is that any covariance matrix $\mathbf{R}$ which meets with equality the partial sum bounds derived using equation (3.4), also minimizes GSC and

---

[1]This example was provided by P. Anigstein at UCB.

maximizes sum capacity [54, 78]. This optimization is a result of the *Schur-convexity* of GSC and C_s [4, 78, 79] in the eigenvalues $\{\lambda\}$ of $\mathbf{R}$, and that the lower bound on partial sums corresponds to a set of eigenvalues which is *majorized* [34] by all other possible eigenvalue sets. The optimization can also be approached using more common (but slightly more cumbersome) stochastic ordering techniques [54]. The only remaining issue is proving the existence of codeword sets which meet the partial eigenvalue sum bounds, and this proof was provided in [78, 79].

Thus, minimizing GSC and maximizing sum capacity are completely equivalent problems for multiple users at one receiver and there exist optimal codeword ensembles which attain maximum C_s and minimum GSC. Some special results are particularly worthy of mention.

- When all users have equal power and there are at least as many users as codeword dimensions, the optimum solution is waterfilling [15, 54, 55, 78, 79] where the codeword covariance waterfills its energy over the ambient noise covariance.

- An "oversized" user [54, 78], for whom the bound of equation (3.4) is satisfied by the middle term and $k = 1$, will reside alone in the signal space dimension with minimum noise. The same goes for users which are oversized with respect to the remaining users and noise dimensions – a sort of *might makes right* statement when it comes to communications resource optimization. The concept will later prove useful when we consider dispersive channels and interference avoidance.

An example of an optimal spectrum which displays a combination of might makes right and waterfilling is shown in FIGURE 3.2

2.   Ensemble Fixed Point Structure

As a prelude to considering codeword ensemble convergence, we first consider the fixed points of interference avoidance algorithms. Specifically, it is clear that no improvement can be had if $\mathbf{s}_k$ is already a minimum eigenvector of $\mathbf{R}_k$ since both $\mathbf{s}_k^\top \mathbf{R}_k \mathbf{s}_k$ would be minimized and $\mathbf{s}_k^\top \mathbf{R}_k^{-1} \mathbf{s}_k$ would be maximized. Thus, if every codeword $\mathbf{s}_k$ is a minimum eigenvector of its corresponding interference covariance matrix $\mathbf{R}_k$, then that ensemble is a *fixed point* of all the distributed interference avoidance algorithms we have presented. For the MMSE algorithm, we note additionally that all the $\mathbf{s}_k$ being ANY eigenvector of $\mathbf{R}_k$ constitutes a fixed point since the replacement codeword is the normalized MMSE filter $\mathbf{c}_k$ and this must be identical to $\mathbf{s}_k$ by equation (2.37).

Regardless, even if all codewords are minimum eigenvectors of their respective interference covariance matrices, minimum GSC or maximum sum capacity is not guaranteed. For instance, all codewords might not necessarily

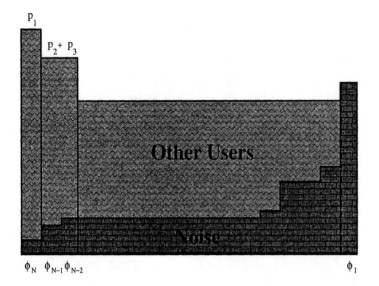

Figure 3.2. Illustration of optimal covariance spectral decomposition for a codeword ensemble with different powers. Vertical bars denote noise covariance dimension partitions.

share the same eigenvalue. That is, there can be different *classes* of users which share different SINRs. Details can be found in [4, 54, 55] and an example is illustrated in FIGURE 3.3, but for our purposes it suffices to say that interference avoidance as thus far presented is not guaranteed to converge to minimum GSC or maximum sum capacity. Of course, this statement begs the questions of whether the fixed points are even reached.

To be more quantitative, consider that at an algorithm fixed point we must have

$$\mathbf{R}_k \mathbf{s}_k = \omega_k \mathbf{s}_k \qquad (3.5)$$

where for the eigen-algorithm, ω_k is a minimum eigenvalue of $\mathbf{R}_k$. We can rewrite this expression in terms of the total covariance $\mathbf{R}$ as

$$\mathbf{R}\mathbf{s}_k = (\omega_k + p_k)\mathbf{s}_k \qquad (3.6)$$

Since $\mathbf{R}$ can have at most N eigenvalues, if there are more users than dimensions, then some users must overlap in the eigenspace of $\mathbf{R}$. However, users which have different values of $\omega_k + p_k$ must be orthogonal since eigenvectors of a symmetric matrix with different eigenvalues are orthogonal. In this way we define a user *class* I as users which share the same covariance eigenvalue $\lambda_I = \omega_k + p_k$, $k \in \mathcal{C}_I$ and therefore can overlap in the eigenspace. It is easy to show that if all users in a given class share the same power, then they also share

Figure 3.3.   Illustration of suboptimal fixed point for a codeword ensemble with different powers.

the same SINR which is in some ways a more practical definition of user *class*. This separation into classes and class orthogonality leads to a useful theorem [55] which states that the each class is spanned by disjoint subsets of noise covariance, $\mathbf{W}$, eigenvectors. An equivalent and equally useful statement is that at any fixed point of the algorithm, the matrices $\mathbf{W}$ and $\mathbf{SS}^\top$ commute [4]. This implied decomposition will prove useful in later chapters.

3.   Convergence

Convergence of interference avoidance, even to potentially suboptimal codewords, has been difficult to prove in general. We first show that greedy interference avoidance [54] converges *in class*. That is, greedy interference avoidance results in sets of codewords which are all eigenvectors of $\mathbf{R}$, but distinguished by possibly different eigenvalues (classes).

THEOREM 3.1 **Convergence In Class for Greedy IA:** *Iterative application of greedy interference avoidance causes codewords to converge to ensembles where each codeword is an eigenvector of the signal plus noise covariance matrix* $\mathbf{R} = \mathbf{R}_i + \mathbf{s}_i\mathbf{s}_i^\top$.

Proof: First we define an *iteration* ℓ as a greedy interference avoidance step where a codeword $\mathbf{s}_{i_\ell}(\ell)$ is replaced. We then have

$$\delta_{i_\ell}(\ell) = \mathbf{s}_{i_\ell}^\top(\ell)\mathbf{R}_{i_\ell}(\ell)\mathbf{s}_{i_\ell}(\ell) - \mathbf{x}_{i_\ell}(\ell)^\top\mathbf{R}_{i_\ell}(\ell)\mathbf{x}_{i_\ell}(\ell) \geq 0 \qquad (3.7)$$

as the difference between the GSC value at iteration ℓ and $\ell + 1$ where $\mathbf{x}_{i_\ell}(\ell)$ is a minimum eigenvalue eigenvector of $\mathbf{R}_{i_\ell}(\ell)$ [55].

We then note that all possible replacement sequences will eventually converge to some GSC value so that we must have

$$\lim_{\ell\to\infty} \delta_{i_\ell} = 0 \qquad (3.8)$$

We define the set of possible iteration sequences as $\{(i_\ell)_m\}$ and note that because each $\delta_{(i_\ell)_m}(\ell)$ must approach zero, given some $\kappa > 0$, there exists an integer $\mathcal{L}_m < \infty$ for each sequence $(i_\ell)_m$ such that

$$\delta_{(i_\ell)_m}(\ell) \leq \kappa \qquad (3.9)$$

where $\ell > \mathcal{L}_m$.

We then consider as $i_\ell = (i_\ell)_{m^*}$ that sequence for which

$$m^* = \operatorname{argmax}_m \mathcal{L}_m \qquad (3.10)$$

Thus, for a given δ-bound κ we consider the codeword replacement sequence $i_\ell = (i_\ell)_{m^*}$ which has largest $\mathcal{L} = \mathcal{L}_{m^*}$ so that κ overbounds $\delta_{(i_\ell)_m}$ for $\ell > \mathcal{L}$ for *all possible codeword adjustment sequences* $(i_\ell)_m$.

Now given κ (and thus i_ℓ and $\mathcal{L}$), consider that the difference in GSC values before and after the *potential* replacement of *any* codeword $\mathbf{s}_i$ at iteration $\ell > \mathcal{L}$ can be written as

$$\kappa \geq \delta_i(\ell) = \mathbf{s}_i^\top(\ell)\mathbf{R}_i(\ell)\mathbf{s}_i(\ell) - \mathbf{x}_i(\ell)^\top\mathbf{R}_i(\ell)\mathbf{x}_i(\ell) \geq 0 \qquad (3.11)$$

We define the eigenvalues of $\mathbf{R}_i(\ell)$ as $\{\lambda_{ij}(\ell)\}, j = 1, 2, \cdots, N$ and assume that they are ordered from largest to smallest. If we further define the corresponding eigevectors as $\phi_{ij}(\ell), j = 1, 2, \cdots, N$ we can rewrite $\mathbf{s}_i(\ell)$ as

$$\mathbf{s}_i(\ell) = \sum_{j=1}^{N} \alpha_{ij}(\ell)\phi_{ij}(\ell) \qquad (3.12)$$

where we assume

$$\sum_{j=1}^{N} \alpha_{ij}^2(\ell) = |\mathbf{x}_i(\ell)|^2 = p_i \qquad (3.13)$$

This leads to

$$\delta_i(\ell) = \sum_{j=1}^{N} \alpha_{ij}^2(\ell)(\lambda_{ij}(\ell) - \lambda_{iN}(\ell)) \qquad (3.14)$$

Since all terms in the sum are non-negative we must have

$$\delta_i(\ell) \geq \alpha_{ij}^2(\ell)(\lambda_{ij}(\ell) - \lambda_{iN}(\ell)) \tag{3.15}$$

for $j = 1, 2, \cdots, N$. Now suppose via equation (3.15) we define $\epsilon_{ij}(\ell) \leq \delta_i(\ell)$ as

$$\epsilon_{ij}(\ell) = \alpha_{ij}^2(\ell)(\lambda_{ij}(\ell) - \lambda_{iN}(\ell)) \tag{3.16}$$

Dividing by nonzero $\alpha_{ij}(\ell)$ results in

$$\frac{\epsilon_{ij}(\ell)}{\alpha_{ij}(\ell)} = \alpha_{ij}(\ell)\lambda_{ij}(\ell) - \alpha_{ij}(\ell)\lambda_{iN}(\ell) \tag{3.17}$$

To see how closely each $s_i(\ell)$ approximates an eigenvector of $R(\ell) = R_i(\ell) + s_i(\ell)s_i^\top(\ell)$ the signal plus noise covariance matrix at iteration ℓ, we form the product

$$R(\ell)s_i(\ell) = \sum_{j \in J_i(\ell)} \lambda_{ij}(\ell)\alpha_{ij}(\ell)\phi_{ij}(\ell) + p_i \sum_{j \in J_i(\ell)} \alpha_{ij}(\ell)\phi_{ij}(\ell) \tag{3.18}$$

where $J_i(\ell)$ is the set of all j such that $\alpha_{ij}(\ell) \neq 0$. Using equation (3.17) in equation (3.18) yields

$$R(\ell)s_i(\ell) = \sum_{j \in J_i(\ell)} \left(\frac{\epsilon_{ij}(\ell)}{\alpha_{ij}(\ell)} + \lambda_{iN}(\ell)\alpha_{ij}(\ell) \right) \phi_{ij}(\ell) + p_i \sum_{j \in J_i(\ell)} \alpha_{ij}(\ell)\phi_{ij}(\ell) \tag{3.19}$$

Regrouping we have

$$R(\ell)s_i(\ell) = (\lambda_{iN}(\ell) + p_i)s_i(\ell) + \sum_{j \in J_i(\ell)} \frac{\epsilon_{ij}(\ell)}{\alpha_{ij}(\ell)}\phi_{ij}(\ell) \tag{3.20}$$

However, since $0 \leq \delta_i(\ell) < \kappa$ and $0 < \alpha_{ij}^2(\ell) < p_i$, then for any $\alpha_{ij}(\ell) \neq 0$ we must have by the definition of $\alpha_{ij}(\ell)$ and equation (3.16)

$$0 \leq \lambda_{ij}(\ell) - \lambda_{iN}(\ell) \leq \frac{\kappa}{\alpha_{ij}^2(\ell)} \tag{3.21}$$

Therefore, we have

$$\left| \frac{\epsilon_{ij}(\ell)}{\alpha_{ij}(\ell)} \right| = |\alpha_{ij}(\ell)| \left(\lambda_{ij}(\ell) - \lambda_{iN}(\ell) \right) \leq \frac{\kappa}{|\alpha_{ij}(\ell)|} \tag{3.22}$$

for any $\alpha_{ij}(\ell)$ *which does not approach zero*. Since $\alpha_{ij}(\ell)$ does not approach zero we have $|\alpha_{ij}(\ell)| > \iota > 0$ which results in

$$\left| \frac{\epsilon_{ij}(\ell)}{\alpha_{ij}(\ell)} \right| \leq \frac{\kappa}{\iota} \tag{3.23}$$

Finally, since we can choose κ arbitrarily small, we then have $\delta_i(\ell)$ and thence $\left|\frac{\epsilon_{ij}}{\alpha_{ij}(\ell)}\right|$ arbitrarily small *for any potential replacement.* This implies that for suitably large ℓ, all codewords $\mathbf{s}_i(\ell)$ are arbitrarily close to being eigenvectors of $\mathbf{R}(\ell)$, thus completing the proof.

We define such codeword convergence as *convergence in class.* ∎

Of course, the theorem is mute on the relative values of the $\lambda_{iN}(\ell) + p_i$ which define classes. Thus, greedy interference avoidance might not result in codewords which produce an optimal covariance matrix $\mathbf{R}$ as defined in section 1, but could theoretically become trapped in local GSC minima corresponding to a suboptimal set of classes.

However, in [4] it was shown that *only* optimal GSC fixed points are stable. That is, only about the optimal fixed point can we find a neighborhood where all GSC values within the neighborhood are greater than or equal to the GSC at the fixed point [4]. Thus, random perturbation of any suboptimal fixed point codeword ensemble has nonzero probability of reducing GSC. Taken together we can either form a modified stochastic algorithm which is guaranteed to converge to the optimal GSC value with probability 1, as was done formally for the MMSE algorithm in [4]. Or we can argue somewhat less formally that starting from random codeword ensembles makes the probability of stopping in a provably unstable suboptimal fixed point identically zero for just about any conceivable interference avoidance algorithm, including the previously introduced lagged IA and gradient descent IA. Within this context it is not surprising that never has interference avoidance been observed to reach a suboptimal point when started from a random ensemble [40, 41, 42, 54, 55, 71].

4. Eigen-Algorithm: Numerical Examples

We now provide simple examples of interference avoidance algorithm application and the resulting codeword sets. First, an optimal signal constellation is shown in FIGURE 3.4 for three codewords in a signal space of dimension 2. FIGURE 3.5 shows codewords for five equal power users with codewords of dimension 3 after five interference avoidance cycles. This speed of convergence is typical and seems to hold for larger number of codewords in higher dimensions as well. [54, 55, 71]. In this case, the resultant associated $\mathbf{SS}^\top$ is approximately diagonal

$$\mathbf{SS}^\top = \begin{bmatrix} 1.667 & -0.00035 & -0.00002 \\ -0.00035 & 1.666 & 0.00017 \\ -0.00002 & 0.00017 & 1.666 \end{bmatrix}$$

and each unit energy user achieves a signal to interference ratio of approximately $\frac{N}{L-N} = 1.5$.

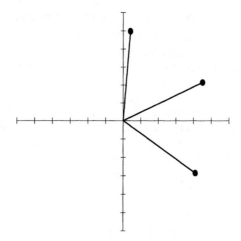

Figure 3.4.   Simple optimal signal set in two dimensions with three users.

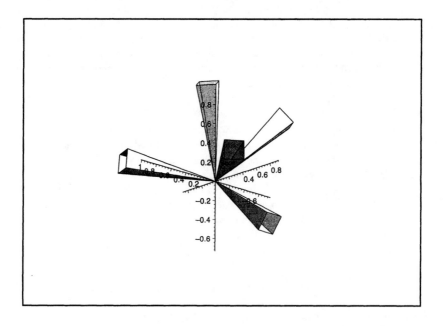

Figure 3.5.   Vector plot of 5 signal vectors in 3-space after 5 cycles of interference avoidance. Signal vectors are represented as hollow inverted pyramids for greater clarity of the 3-dimensional representation.

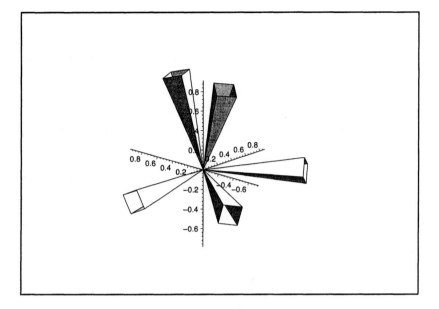

Figure 3.6. Vector plot of 5 signal vectors (3 agile and 2 fixed) after 5 cycles of interference avoidance. Notice the co-planarity of the three agile signal vectors in avoidance of a strong fixed interferer component along the remaining dimension.

In FIGURE 3.6 we allow only the first three users to perform interference avoidance and fix two others which effectively act as colored noise. After 5 cycles the three *agile* user codewords reside in a space of dimension 2 (coplanar) achieving approximately $\lambda^* = 1.6$ and a concomitant signal to interference ratio of approximately $(\lambda^* - 1)^{-1} = 1.\overline{66}$. They avoid a strong fixed user interference component (with energy 1.8) in the remaining dimension.

In FIGURE 3.7 we assume four users, one of whom has much larger codeword power than the others ($p_1 = 4$). This energetic user commands a private channel and the remaining three users are forced to share two dimensions of the signal space – again the weaker users are coplanar – and have $\lambda^* = 3/2$ for a shared SIR of 2.

5. Summary

In this chapter we considered some of the details regarding the structure of optimal codeword ensembles and thus derived lower bounds on GSC and upper bounds on sum capacity. We also considered convergence of codeword ensembles under interference avoidance. We found that often the optimal solution is a water filling solution which by definition maximizes sum capacity. We also

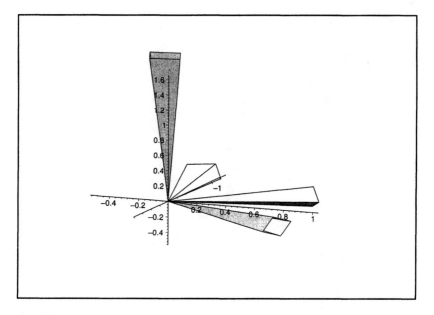

Figure 3.7.   Vector plot of 4 signal vectors, three with power 1 and one with power 4. Notice the co-planarity of the three power 1 vectors in avoidance of the strong interferer with power 4.

learned that when there exist users with power much greater than other users (called an *oversized user*), that user's "might makes right" and it will reside alone in the best (lowest noise) dimension of the signal space. We also offered quantitative and qualitative explanations for the empirically seen convergence of various interference avoidance algorithms, including discussion of a formal convergence proof for the MMSE version of interference avoidance. We then closed by considering simple examples in visualizable signal spaces to illustrate the basic principles.

Chapter 4

INTERFERENCE AVOIDANCE ALGORITHMS FOR NON-IDEAL CHANNELS

In the previous chapters we assumed that communication channels were characterized by an ideal impulse response, and that the only impairment consisted of additive Gaussian noise. However, communications channels are in general dispersive, and dispersion leads to intersymbol interference (ISI) where the energy of a given symbol spills over into the observation intervals of adjacent symbols at the receiver. This chapter extends application of interference avoidance methods to dispersive multiple access channels.

Over the years, two main approaches have been taken in dealing with ISI. The traditional approach was to use channel equalization [48, Ch. 10] and coding [83] techniques which compensate for channel distortion and correct potential transmission errors. Over the past fifteen years or so, multicarrier modulation has emerged as a viable alternative for high speed data transmission systems [5]. The idea behind multicarrier modulation is to employ block transmissions in frames of extended duration. In this case, a relatively small "zero pad" between successive blocks allows for settling of channel impulse responses and makes ISI irrelevant. The concept is quite old and its theoretical origins date back almost 40 years to the work of Holsinger [25].

More recent work performed in the area of multicarrier modulation can be grouped into two main camps: optimization of transmitted power, and multiple access techniques and signal design. A framework for using multicarrier modulation in frequency dispersive multiple access channels based on discrete multi-tone (DMT) schemes is presented in [16], where a multiuser bit-loading algorithm is also proposed. The algorithm performs multiuser water filling [10] distribution over the DMT tones and maximizes the sum capacity of the multi-access channel. The fact that a DMT scheme with appropriately loaded carriers is optimal with respect to maximizing sum capacity subject to a given power constraint has been proved to be optimal [38]. A general iterative water filling

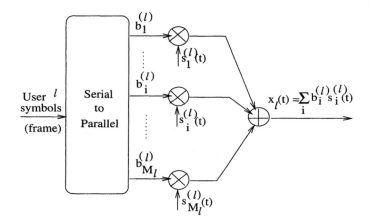

Figure 4.1. Multicode CDMA approach for sending frames of information. Each symbol in the frame is assigned a signature waveform and the resulting signal $x_\ell(t)$ is a superposition of signature waveforms scaled by their corresponding information symbols.

procedure applicable to multiaccess vector channel models has been proposed proposed recently [86].

In the area of signal design we note the work of Honig et al. [26] where optimum signal sets for dispersive channels are derived in a framework based on channel eigen-decomposition [25]. We also note the work of Kasturia et. al [29] and Lechleider [32] which propose codeword design methods for block transmissions suited for multicarrier modulation systems. More recently, in the context of multiuser detection, methods for transmitter and receiver adaptation [53] have also been used for non-ideal channels, although not for a multicarrier modulation framework in particular. Here we note the work of Rajappan and Honig [51], and that of Concha and Ulukus [12]. In this case transmitter/receiver adaptation compensates for the distortion introduced by the channel and avoids multiaccess interference.

Our approach to extending interference avoidance methods to dispersive channels uses a multicarrier modulation framework, and falls in the second category, that of signal design. Specifically, we consider the uplink of a synchronous CDMA system with L users, where each user sends frames containing multiple symbols using a multicode CDMA approach as described schematically in Figure 4.1. Each symbol in a given user's frame is assigned a specific signature waveform and the transmitted signal is

$$x_\ell(t) = \sum_{m=1}^{M_\ell} b_m^{(\ell)} s_m^{(\ell)}(t) \qquad \ell = 1, \ldots, L \tag{4.1}$$

where $b_m^{(\ell)}$, $m = 1, \ldots, M_\ell$, denote the symbols sent by user ℓ, and $s_m^{(\ell)}(t)$ is the signature waveform assigned to convey symbol m of user ℓ assumed of finite duration T.

The channel between a given user ℓ and the base station is assumed to be linear and time-invariant and is characterized by the causal impulse response $h_\ell(t)$ of duration T_ℓ. We assume that the frame duration is much larger than the duration of all channel impulse responses $T \gg T_\ell$, $\forall \ell = 1, \ldots, L$. Thus, one can safely ignore ISI between successive frames of duration T by placing a relatively small "zero pad" between them to allow settling of channel responses.

The received signal at the base station is a sum of signals transmitted by all users convolved with their corresponding impulse responses plus additive Gaussian noise $n(t)$

$$r(t) = \sum_{\ell=1}^{L} x_\ell(t) * h_\ell(t) + n(t) \qquad (4.2)$$

Our goal is to apply interference avoidance to derive optimal signature waveform ensembles $\{s_m^{(\ell)}(t)\}$, $\ell = 1, \ldots, L$, $m = 1, \ldots, M_\ell$ that maximize the sum capacity of the multiaccess dispersive channel in the uplink of a CDMA system. In order to do this we will first convert the continuous-time (waveform) multiaccess channel in equation (4.2) into an equivalent vector multiaccess channel. We note that a vector channel is a natural representation of a waveform channel in a finite dimensional signal space [20, Ch. 8] implied by finite frame duration and finite bandwidth.

For a linear and time-invariant channel characterized by impulse response $h(t)$ a useful set of basis functions which allows convenient representation of channel inputs and outputs are the eigenfunctions of the channel impulse response *autocorrelation function* introduced by Holsinger in [25]. This set of functions forms an orthonormal basis for the signal space and can be used to represent channel inputs as vectors in an Euclidian space of the same dimension as the signal space, and hence to convert the waveform channel into an equivalent vector channel as shown in Figure 4.2. A detailed derivation of channel eigenfunctions is beyond the scope of this book, and we refer interested readers to the original work by Holsinger [25], or a more standard textbook [20, Sec. 8.4]. We note that, while in general channel eigenfunctions have complicated mathematical expressions and differ from channel to channel, under the assumption that the frame duration T is large relative the durations of all the channel impulse responses, channel eigenfunctions will all be approximately sinusoidal, thus allowing the use of the same orthonormal basis for all users to convert waveform channels into their equivalent vector channels.

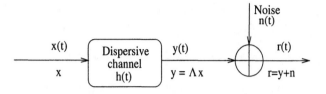

Figure 4.2. Equivalent descriptions of the same dispersive channel using the waveform and vector channel representations.

In this framework, the waveform channel in equation (4.2) is equivalent to the multiaccess vector channel

$$\mathbf{r} = \sum_{\ell=1}^{L} \mathbf{\Lambda}_\ell \mathbf{x}_\ell + \mathbf{n} \qquad (4.3)$$

where $\mathbf{x}_\ell$, denotes the signal corresponding to user ℓ, $\mathbf{n}$ is the additive Gaussian noise vector that corrupts the signal at the base station receiver, and $\mathbf{\Lambda}_\ell$ is the matrix that describes the uplink channel for user ℓ. This is a standard vector channel model commonly used in conjunction with multicarrier modulation systems, and its derivation can be found in any standard communications textbook (see for example [23, Sec. 6.12-13]).

This model corresponds to a signal space representation in terms of a finite number of frequencies for which the channel matrix $\mathbf{\Lambda}_\ell$ of a given user ℓ is diagonal and contains the channel gains corresponding to frequencies that span the signal space. We note that the real (double-sided) representation, in which sine and cosine of the same frequency make up two orthogonal signal space dimensions with the same real channel gain, implies a vector channel of dimension $2N$, where N is the number of frequencies that span the signal space. Alternatively, the complex (single-sided) representation favored in the multicarrier and OFDM literature, which uses complex exponentials as basis functions, implies a vector channel of dimension N with complex gains corresponding to each of the N frequencies that span the signal space.

Since each user employs multicode CDMA to transmit frames containing multiple symbols the transmitted signal vector for user ℓ is

$$\mathbf{x}_\ell = \sum_{m=1}^{M_\ell} b_m^{(\ell)} \mathbf{s}_m^{(\ell)} = \mathbf{S}_\ell \mathbf{b}_\ell \qquad \ell = 1, \ldots, L \qquad (4.4)$$

where $\mathbf{s}_m^{(\ell)}$ is the codeword (signature sequence) that corresponds to waveform $s_m^{(\ell)}(t)$.

If the real representation is used then the transmitted symbols and their corresponding codewords are also real, while in the case of the complex representation transmitted symbols and codewords will be complex. We note that if at least as many symbols as signal space dimensions are transmitted by a given user ℓ, then its transmit covariance matrix $\mathbf{X}_\ell = \mathbf{S}_\ell \mathbf{S}_\ell^\top$ may span all available signal space dimensions. We also note that the isomorphism between the field of complex numbers and the field of 2×2 real and skew-symmetric matrices guarantees complete equivalence between the real and complex representations. For simplicity of the mathematical presentation and with no loss of generality, we choose to work with the real framework corresponding to the double-sided signal representation.

To summarize, the received signal vector is expressed as

$$\mathbf{r} = \sum_{\ell=1}^{L} \mathbf{\Lambda}_\ell \mathbf{S}_\ell \mathbf{b}_\ell + \mathbf{n} \qquad (4.5)$$

and our goal is to derive optimal ensembles of codewords, or signature sequences, $\{\mathbf{s}_m^{(\ell)}\}$, $\ell = 1, \ldots, L$, $m = 1, \ldots, M_\ell$ which maximize the sum capacity of the dispersive multiaccess vector channel in equation (4.5). We note that this is different from the work of Yu et al. [86] which deals essentially with optimal power allocation. While in [86] Yu et al. look for the set of optimal transmit covariance matrices $\mathbf{X}_\ell$ which maximize sum capacity and present a water filling procedure, our goal is to provide a codeword adaptation algorithm and not a water filling procedure. Nevertheless, since we are looking at the same performance criterion as Yu et al. in [86], namely sum capacity, the structure of the transmit covariance matrices implied by the optimal codeword ensemble $\mathbf{X}_\ell = \mathbf{S}_\ell \mathbf{S}_\ell^\top$ will satisfy a similar water filling solution.

1. The Single User Case

Application of greedy interference avoidance for a single user that communicates over a dispersive channel is a straightforward application of the eigen-algorithm which was presented in Chapter 2.

In the case of a single user, the received signal in equation (4.5) becomes

$$\mathbf{r} = \mathbf{\Lambda} \mathbf{S} \mathbf{b} + \mathbf{n} \qquad (4.6)$$

Assuming that $\mathbf{\Lambda}$ is invertible, we can rewrite equation (4.6) as

$$\tilde{\mathbf{r}} = \mathbf{\Lambda}^{-1} \mathbf{r} = \mathbf{S} \mathbf{b} + \tilde{\mathbf{n}} \qquad (4.7)$$

in which $\tilde{\mathbf{n}} = \mathbf{\Lambda}^{-1} \mathbf{n}$ is a new vector of noise. Equation (4.7) which describes the equivalent problem is identical to equation (2.4) and allows straightforward application of the eigen-algorithm in Chapter 2 to determine the optimal codeword ensemble.

We note that the assumption of channel invertibility is not a restriction in the context of the water filling solution implied by the eigen-algorithm for interference avoidance. This is because water filling solutions dictate that if the noise energy in a dimension is large enough relative other dimensions, then no signal energy can reside in that dimension. Thus, in the context of equation (4.7), the optimal codeword ensembles will be identical for non-invertible channels and their counterparts made invertible by replacing zero gain elements by sufficiently small but nonzero gains. A careful definition of "sufficiently small" can be stated as a theorem whose proof is a simple consequence of water filling.

THEOREM 4.1 *Following equation (4.6), let us consider a non-invertible channel gain matrix Λ^* with $k > 0$ nonzero gains and the invertible matrix Λ*

$$\Lambda^* = \begin{bmatrix} \lambda_1 & 0 & \cdots & \cdots & \cdots & \cdots & 0 \\ 0 & \ddots & 0 & \cdots & \cdots & \cdots & 0 \\ \vdots & 0 & \lambda_k & 0 & \cdots & \cdots & \vdots \\ \vdots & \cdots & 0 & 0 & \ddots & \cdots & \vdots \\ \vdots & \cdots & \cdots & \ddots & \ddots & \ddots & \vdots \\ \vdots & \cdots & \cdots & \cdots & \ddots & \ddots & 0 \\ 0 & \cdots & \cdots & \cdots & \cdots & 0 & 0 \end{bmatrix} \quad \Lambda = \begin{bmatrix} \lambda_1 & 0 & \cdots & \cdots & \cdots & \cdots & 0 \\ 0 & \ddots & 0 & \cdots & \cdots & \cdots & 0 \\ \vdots & 0 & \lambda_k & 0 & \cdots & \cdots & \vdots \\ \vdots & \cdots & 0 & \epsilon & \ddots & \cdots & \vdots \\ \vdots & \cdots & \cdots & \ddots & \ddots & \ddots & \vdots \\ \vdots & \cdots & \cdots & \cdots & \ddots & \ddots & 0 \\ 0 & \cdots & \cdots & \cdots & \cdots & 0 & \epsilon \end{bmatrix}$$

and assume with no loss of generality that $\lambda_i \geq \lambda_{i+1}$. Likewise consider a diagonal noise covariance matrix

$$E[\mathbf{nn}^\mathsf{T}] = \mathbf{W} = \begin{bmatrix} \sigma_1^2 & 0 & \cdots & 0 \\ 0 & \ddots & \ddots & \vdots \\ \vdots & \ddots & \ddots & 0 \\ 0 & \cdots & 0 & \sigma_N^2 \end{bmatrix}$$

Finally assume unit energy symbols b_i so that the total transmitted signal energy over all dimensions is $E = Trace\left[\mathbf{SS}^\mathsf{T}\right]$. If ϵ is chosen such that

$$\epsilon < \frac{\sigma_j^2}{\dfrac{1}{k}\left[E + \displaystyle\sum_{i=1}^{k} \dfrac{\sigma_i^2}{\lambda_i}\right]}$$

$j = k + 1, \ldots, N$, then the set of codeword ensembles which maximize sum capacity for the channel of equation (4.6) will be identical for Λ^* and Λ.

Proof: The theorem is a simple consequence of water filling and of the fact that interference avoidance provides a codeword ensemble which water fills the signal space. ∎

Therefore, in what follows we assume all channels are invertible with no loss of generality.

2. The Multiuser Case

In this section we consider the general case with multiple users for which the received signal is described by equation (4.5). From the perspective of user k equation (4.5) can be rewritten as

$$\mathbf{r} = \Lambda_k \mathbf{S}_k \mathbf{b}_k + \sum_{\ell=1,\ell\neq k}^{L} \Lambda_\ell \mathbf{S}_\ell \mathbf{b}_\ell + \mathbf{n} \tag{4.8}$$

in which the first term is the desired signal corresponding to user k while the rest represents interference coming from other users and noise. We note that all the Λ_ℓ matrices are assumed invertible, although some of their elements may be of $O(\varepsilon)$. Nevertheless, as pointed out in section 1 via Theorem 4.1, this does not restrict application of greedy interference avoidance since those dimensions corresponding to very small gains will be completely avoided.

Assuming that noise is colored with uncorrelated components, the covariance matrix of the received signal is

$$\mathbf{R} = E[\mathbf{r}\mathbf{r}^\top] = \sum_{\ell=1}^{L} \Lambda_\ell \mathbf{S}_\ell \mathbf{S}_\ell^\top \Lambda_\ell + \mathbf{W} \tag{4.9}$$

with $\mathbf{W} = E[\mathbf{n}\mathbf{n}^\top]$ a diagonal matrix with elements equal to σ_i^2, $i = 1 \ldots N$, representing the noise variances along each signal space dimension.

From the perspective of an individual user, our problem is again that of selecting input codewords for its symbols in the presence of combined noise and interference from other users. Similar to equation (4.7) we define an equivalent problem for user k, pre-multiplying by the corresponding inverse channel matrix Λ_k^{-1} in equation (4.8) to obtain

$$\mathbf{r}_k = \mathbf{S}_k \mathbf{b}_k + \Lambda_k^{-1} \left(\sum_{\ell\neq k} \Lambda_\ell \mathbf{S}_\ell \mathbf{b}_\ell + \mathbf{n} \right) \tag{4.10}$$

The covariance matrix of the received signal corresponding to user k inverted channel problem is

$$\mathbf{R}^{(k)} = \mathbf{S}_k \mathbf{S}_k^\top + \Lambda_k^{-1} \left(\sum_{\ell\neq k} \Lambda_\ell \mathbf{S}_\ell \mathbf{S}_\ell^\top \Lambda_\ell + \mathbf{W} \right) \Lambda_k^{-1} \tag{4.11}$$

and is related to the original received signal covariance matrix by

$$\mathbf{R}^{(k)} = \Lambda_k^{-1} \mathbf{R} \Lambda_k^{-1} \tag{4.12}$$

The greedy interference avoidance procedure can be applied now for user k equivalent problem by replacing codeword m of user k, $\mathbf{s}_m^{(k)}$, with the minimum eigenvector of the corresponding interference-plus-noise covariance matrix under channel k inversion given by

$$\mathbf{R}_m^{(k)} = \mathbf{R}^{(k)} - \mathbf{s}_m^{(k)} \mathbf{s}_m^{(k)\top} \tag{4.13}$$

This leads to a straightforward generalization of the eigen-algorithm in the multiuser case which is formally stated below:

The Multiuser Eigen-Algorithm for Dispersive Channels

1 Start with a randomly chosen codeword ensemble specified by the user codeword matrices $\mathbf{S}_1, \ldots, \mathbf{S}_L$

2 For each user $k = 1 \ldots L$

   (a) Define the equivalent problem for user k as in equation (4.10)

   (b) adjust user k's codewords sequentially: the codeword corresponding to symbol m of user k is replaced by the minimum eigenvector of the auto-correlation matrix of the corresponding interference-plus-noise process in equation (4.13)

   (c) Repeat step (b) iteratively for each user until a fixed point is reached for which further modification of codewords will bring no additional improvement.

   (d) If a suboptimal point is reached use escape methods [54] and repeat steps (b)-(c).

3 Repeat step 2.

We will show that fixed points of the multiuser eigen-algorithm for dispersive channels exist and yield codeword ensembles which maximize sum capacity. We first note that application of greedy interference avoidance in the multiuser/multicode CDMA context monotonically increases sum capacity, which is given in this case by an expression identical to that in equation (2.14), but in which $\mathbf{R}$ has the more complex expression in equation (4.9). Using the relationship in equation (4.12) we rewrite sum capacity from user k perspective

$$
\begin{aligned}
C_s &= \frac{1}{2} \log \left[\det(\mathbf{\Lambda}_k \mathbf{R}^{(k)} \mathbf{\Lambda}_k) \right] - \frac{1}{2} \log(\det \mathbf{W}) \\
&= \frac{1}{2} \log(\det \mathbf{R}^{(k)}) + \frac{1}{2} \log(\det \mathbf{\Lambda}_k^2) - \frac{1}{2} \log(\det \mathbf{W})
\end{aligned}
\tag{4.14}
$$

Using the results from Chapter 2 we note that the last two terms are constant while user k applies greedy interference avoidance. Furthermore, from equation

(4.13) we have that $\mathbf{R}^{(k)} = \mathbf{R}_m^{(k)} + \mathbf{s}_m^{(k)} \mathbf{s}_m^{(k)^\top}$ and following the same line of reasoning as in Chapter 2 we get the desired result.

Since application of greedy interference avoidance at each step of the eigen-algorithm monotonically increases sum capacity, and because sum capacity is upper bounded, we can conclude that the multiuser eigen-algorithm converges to some fixed point. We do not yet claim that this corresponds to the maximum sum capacity point, just that convergence to a fixed point is certain. In fact, we will present later on a looser application of greedy interference avoidance in which codewords are updated randomly over all users (as opposed to updating a single user's codewords until convergence and then moving on) and which, by the same token, is also guaranteed to converge to a fixed point.

Next, we recall that application of the eigen-algorithm for a system with more codewords than signal space dimensions ($M_k \geq N$) in a colored noise background performs an *aggregate water filling* over that portion of the signal space with least interference-plus-noise energy, leading to an optimum SINR for all codewords. Therefore, for $M_k \geq N$, in Step 2 of the multiuser eigen-algorithm user k water fills those dimensions of its inverted channel problem with minimum interference plus noise energy, and avoids dimensions with interference plus noise energy above a "watermark". As a consequence, the fixed point of the multiuser eigen-algorithm is characterized by the property that all codewords of any given user k are minimum eigenvectors of the received signal covariance matrix that corresponds to user k inverted channel problem

$$\mathbf{R}^{(k)}\mathbf{S}_k = \mu_k \mathbf{S}_k \quad \forall k = 1, \ldots, L \tag{4.15}$$

The eigenvalue μ_k denotes the "watermark" associated with user k water filling distribution in its equivalent inverted channel problem, and is also related to the uniform SINR achieved by all symbols of user k as $\gamma_k = 1/(\mu_k - 1)$. A graphical illustration of such a water filling solution for a system with two users in a signal space spanned by 6 frequencies is shown in Figure 4.3.

Empirically we have observed that with a diagonal noise covariance $\mathbf{W}$, the received signal covariance matrix $\mathbf{R}$ at the fixed point of the multiuser eigen-algorithm is diagonal. We defer analytical proof of this result until section 3, and note that with diagonal noise covariance matrix a diagonal received signal covariance matrix can be obtained when all codeword covariances $\mathbf{S}_\ell \mathbf{S}_\ell^\top$ for all users $\ell = 1, \ldots, L$ are also diagonal. While this may not be the only codeword covariance structure which could result from application of the algorithm, it is an obvious solution which implies a diagonal structure at a fixed point of the algorithm. In addition, *all* simulations of the eigen-algorithm using diagonal $\mathbf{W}$ have yielded this structure. Furthermore, when users are added to the system one by one while applying the eigen-algorithm, according to the following scenario, we are guaranteed to obtain codeword ensembles with diagonal co-

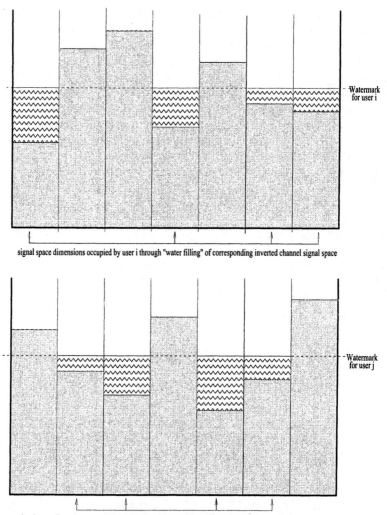

signal space dimensions occupied by user i through "water filling" of corresponding inverted channel signal space

signal space dimensions occupied by user j through "water filling" of corresponding inverted channel signal space

Figure 4.3. Example of Simultaneous Water Filling for a System With Two Users. Due to the Different Interference-Plus-Noise Levels On Different Signal Space Dimensions, Users i and j Span Different (Possibly Overlapping) Subspaces.

variance matrices. We start with all users silent and add them to the system one by one. In the presence of uncorrelated noise with diagonal covariance matrix $\mathbf{W}$, the orthogonal basis used for generating codewords for user 1 through greedy interference avoidance is exactly the canonical basis of $\mathbb{R}^N$ consisting of vectors $\mathbf{e}_i = [0, \ldots, 0, 1, 0, \ldots, 0]^\top$ (zero except for 1 in the ith position), and will yield a codeword ensemble with diagonal covariance matrix for user 1. Since channel gain matrices are also diagonal, the resulting $\mathbf{R}^{(1)}$ will also be diagonal, which in turn implies that when the next user is added to the system, it will also see a diagonal noise-plus-interference covariance matrix in its corresponding inverted channel problem, and so on, for remaining users added one by one into the system. After all users are added, we may begin again with an arbitrarily chosen user. Regardless, the noise/interference structure seen by that user is still diagonal and the chosen user will occupy a subspace spanned by canonical vectors. Thus, sequential application of the eigen-algorithm for interference avoidance will yield at equilibrium codeword ensembles with diagonal covariance matrices.

We are now ready to show that the fixed point of the eigen-algorithm where each user water fills its own inverted channel signal space corresponds to a signal constellation which maximizes sum capacity. To do so, we derive an expression for sum capacity in the context of our formulation and show that it is maximized by any signal set which satisfies the fixed point of the eigen-algorithm.

In order to maximize the sum capacity for the multiple access channel defined by equation (4.5), the eigenvalues of the received signal covariance matrix $\mathbf{R}$ must be solutions of the following optimization problem

$$\max_{\mathbf{s}_m^{(\ell)}, \, m=1,\ldots,M_\ell, \, \ell=1,\ldots,L} C_{\text{sum}} \tag{4.16}$$

subject to some power constraint on the codewords. Since unit energy codewords have been assumed for all users, the power constraint is written as

$$\text{Trace} \left[\mathbf{S}_\ell \mathbf{S}_\ell^\top \right] = M_\ell, \quad \forall \ell = 1, \ldots, L \tag{4.17}$$

We assume that each user transmits at least as many codewords as signal space dimensions, that is $M_\ell \geq N$, which guarantees that users are not restricted to a lower dimensional subspace and may span the entire signal space. We note that the second term in the sum capacity expression in equation (2.14) is fixed and therefore sum capacity is maximized when the first term is maximized, that is when $\det \mathbf{R}$, with $\mathbf{R}$ given by equation (4.9), is maximized. The diagonal covariance matrix structure that corresponds to a fixed point implies that we can write

$$\det \mathbf{R} = \prod_{n=1}^{N} \left(\sum_{\ell=1}^{L} \lambda_n^{(\ell)2} \varrho_n^{(\ell)} + \sigma_n^2 \right) \tag{4.18}$$

with $\varrho_n^{(\ell)}$ being the elements of diagonal codeword covariance matrix

$$
\mathbf{S}_\ell \mathbf{S}_\ell^\top = \begin{bmatrix} \varrho_1^{(\ell)} & & & & \\ & \ddots & & & \\ & & \varrho_n^{(\ell)} & & \\ & & & \ddots & \\ & & & & \varrho_N^{(\ell)} \end{bmatrix} \tag{4.19}
$$

With this notation the constrained optimization problem in equations (4.16)–(4.17) can be restated as

$$
\text{maximize} \quad \frac{1}{2} \sum_{n=1}^{N} \log \left(1 + \frac{1}{\sigma_n^2} \sum_{\ell=1}^{L} \lambda_n^{(\ell)2} \varrho_n^{(\ell)} \right) \tag{4.20}
$$

subject to the constraints

$$
\text{Trace}\left[\mathbf{S}_\ell \mathbf{S}_\ell^\top \right] = \sum_{n=1}^{N} \varrho_n^{(\ell)} = M_\ell, \qquad \forall \ell = 1, \ldots, L \tag{4.21}
$$

and

$$
\varrho_n^{(\ell)} \geq 0, \qquad \forall \ell = 1, \ldots, L, \; n = 1, \ldots, N \tag{4.22}
$$

and we prove the following result:

THEOREM 4.2 *The eigenvalue distribution for the codeword ensemble that maximizes sum capacity corresponds to a water filling solution in the inverted channel signal space for each user.*

Proof: In order to solve the constrained optimization problem defined by equations (4.20)-(4.22) we use the Lagrange multipliers method and construct

$$
J = \frac{1}{2} \sum_{n=1}^{N} \log \left(1 + \frac{1}{\sigma_n^2} \sum_{\ell=1}^{L} \lambda_n^{(\ell)2} \varrho_n^{(\ell)} \right) + \xi_1 \left(\sum_{n=1}^{N} \varrho_n^{(1)} - M_1 \right)
$$
$$
+ \quad \ldots + \xi_L \left(\sum_{n=1}^{N} \varrho_n^{(L)} - M_L \right) \tag{4.23}
$$

and then compute the derivative for user i, dimension n, i.e.

$$
\frac{\partial J}{\partial \varrho_n^{(i)}} = \frac{1}{2} \cdot \frac{1}{1 + \frac{1}{\sigma_n^2} \sum_{\ell=1}^{L} \lambda_n^{(\ell)2} \varrho_n^{(\ell)}} \cdot \frac{\lambda_n^{(i)2}}{\sigma_n^2} + \xi_i \tag{4.24}
$$

According to the Kuhn-Tucker theorem [6, p. 429] the necessary and sufficient conditions for maximizing J are

$$\frac{\partial J}{\partial \varrho_n^{(i)}} \geq 0 \qquad (4.25)$$

with equality if $\varrho_n^{(i)} \neq 0$. For those dimensions n for which $\varrho_n^{(i)} \neq 0$ equation (4.24) can be rewritten as

$$\frac{\lambda_n^{(i)2}}{\sigma_n^2 + \sum_{\ell=1}^{L} \lambda_n^{(\ell)2} \varrho_n^{(\ell)}} + 2\xi_i = 0 \qquad (4.26)$$

By taking all the derivatives for $n = 1, \dots, N$ for user i we get the equalities

$$\frac{\lambda_{n_1}^{(i)2}}{\sigma_{n_1}^2 + \sum_{\ell=1}^{L} \lambda_{n_1}^{(\ell)2} \varrho_{n_1}^{(\ell)}} = \dots = \frac{\lambda_{N_i}^{(i)2}}{\sigma_{N_i}^2 + \sum_{\ell=1}^{L} \lambda_{N_i}^{(\ell)2} \varrho_{N_i}^{(\ell)}} \qquad (4.27)$$

for all those dimensions $n = n_1, \dots, N_i$ corresponding to non-zero $\varrho_n^{(i)}$. This can also be rewritten as

$$\frac{1}{\varrho_{n_1}^{(i)} + \frac{\sigma_{n_1}^2}{\lambda_{n_1}^{(i)2}} + \frac{1}{\lambda_{n_1}^{(i)2}} \sum_{\ell=1,\ell\neq i}^{L} \lambda_{n_1}^{(\ell)2} \varrho_{n_1}^{(\ell)}} = \dots$$

$$\dots = \frac{1}{\varrho_{N_i}^{(i)} + \frac{\sigma_{N_i}^2}{\lambda_{N_i}^{(i)2}} + \frac{1}{\lambda_{N_i}^{(i)2}} \sum_{\ell=1,\ell\neq i}^{L} \lambda_{N_i}^{(\ell)2} \varrho_{N_i}^{(\ell)}} \qquad (4.28)$$

The equalities in equation (4.28) define water filling under channel inversion for user i and they yield the distribution of power along signal space dimensions for user i. Note that equation (4.28) is exactly the classical water filling equation (see for example [15]) for the case when user i has a flat channel, and both noise and interference coming from other users in each signal space dimension is scaled by the corresponding channel gain of user i. ∎

We then note that the expression for capacity is concave in the $\varrho_n^{(i)}$. Thus, any solution which satisfies the extremal conditions of Theorem 4.2 must also be optimal. We conclude that the optimal point with maximum sum capacity corresponds to a simultaneous water filling distribution for users in their equivalent inverted channel problems.

It is worth noting that the solution of the constrained optimization problem in equations (4.16)–(4.17) corresponds in fact to the optimal power distribution for all users along the signal space dimensions for the codeword ensemble that maximizes sum capacity of the given multiaccess dispersive channel. This is so because diagonal elements in $S_\ell S_\ell^\top$ represent the amount of power (energy) put by user ℓ into the corresponding signal space dimension. This comes in agreement with the recent and more general result [86] which show that a *simultaneous water filling* solution attains the largest possible sum capacity. We will refer to this more general result later in Chapter 5 where we will extend application of greedy interference avoidance to multiaccess vector channels. We also note that from the perspective of the general algorithm defined in [86] the multiuser eigen-algorithm is an instance of the iterative water filling procedure.

To conclude the exposition so far, we reiterate the fact that the multiuser eigen-algorithm in which all codewords of a given user are updated sequentially based on greedy interference avoidance until convergence and the procedure is repeated for all users in the system, converges to a simultaneous water filling solution which corresponds to maximum sum capacity. However, as mentioned earlier, we have observed empirically that repeated application of greedy interference avoidance with various codeword replacement orders reaches the same optimal fixed point – unless the procedure is deliberately placed in a suboptimal fixed point at initialization. We need to point out that we do not claim that codewords converge to a particular codeword ensemble, but rather that the procedure converges to a class of codewords which corresponds to maximum sum capacity, the so-called convergence in class mentioned in Chapter 3. We also emphasize that, in general, these codeword adaptation procedures are different from water filling schemes, even though they converge to a water filling solution. While we have been unable to prove convergence to the optimal fixed point in general, we mention an alternative algorithm based on greedy interference avoidance for which convergence to the optimal point is provable. This is formally stated below and at each step the codeword that maximally increases sum capacity over all codewords of all users is selected and replaced with the minimum eigenvector of the corresponding interference-plus-noise covariance matrix.

The Maximum Capacity Increase Algorithm for Interference Avoidance

1 Start with a randomly chosen codeword ensemble specified by the user codeword matrices $S_1, \ldots, S_L$

2 Define the equivalent problems for all users k as in equation (4.10)

3 Identify the codeword $s_m^{(k)}$ whose replacement will maximally increase sum capacity. If no codeword will increase sum capacity, and suboptimal maxima escape methods [54] are ineffective for improvement, then STOP. Otherwise,

    (a) adjust $s_m^{(k)}$: replacement by the minimum eigenvector of the autocorrelation matrix of the corresponding interference-plus-noise process in equation (4.13)

    (b) Return to step 2

First we note that because at each step the maximum capacity increase algorithm is based on a greedy interference avoidance procedure it monotonically increases sum capacity. Furthermore, the maximum capacity increase algorithm stops only if sum capacity cannot be increased. Thus, the sequence of sum capacity values along any update trajectory must be strictly increasing. In order to prove that the algorithm converges to the optimal point, which corresponds to a simultaneous water filling solution for all users in their respective equivalent inverted channel problems, we first show that in the limit, the maximum capacity increase algorithm for interference avoidance produces codewords $s_m^{(k)}$ which are eigenvectors of $\mathbf{R}^{(k)}$, $k = 1, 2, ..., L$, $m = 1, 2, ..., M_k$.

Let k_p be the user index and m_p the codeword index for that user chosen for update at algorithm step p. Let the N eigenvalues $\{\gamma_i\}$ of $\mathbf{R}_{m_p}^{(k_p)}$ be ordered from largest to smallest and let the associated eigenvectors be $\{\phi_i\}$. Then, C_p, the sum capacity after each step p of the algorithm is

$$C_p = \frac{1}{2} \log \left| \mathbf{R}_{m_p}^{(k_p)} + \phi_N \phi_N^\top \right| - \frac{1}{2} \log |\mathbf{W}| + \frac{1}{2} \log |\Lambda_{k_p}| \qquad (4.29)$$

The change in sum capacity, $\Delta_p = C_p - C_{p-1}$ can then be written as

$$\Delta_p = \frac{1}{2} \log \left| \mathbf{R}_{m_p}^{(k_p)} + \phi_N \phi_N^\top \right| - \frac{1}{2} \log \left| \mathbf{R}_{m_p}^{(k_p)} + s_{m_p}^{(k_p)} (s_{m_p}^{(k_p)})^\top \right| > 0 \quad (4.30)$$

which we rewrite as

$$e^{-2\Delta_p} = \frac{\left| \mathbf{R}_{m_p}^{(k_p)} + s_{m_p}^{(k_p)} (s_{m_p}^{(k_p)})^\top \right|}{\left| \mathbf{R}_{m_p}^{(k_p)} + \phi_N \phi_N^\top \right|} \qquad (4.31)$$

Then we note that for

$$\Phi = \begin{bmatrix} | & | & & | \\ \phi_1 & \phi_2 & \cdots & \phi_N \\ | & | & & | \end{bmatrix} \qquad (4.32)$$

and

$$\mathbf{\Gamma} = \begin{bmatrix} \gamma_1 & 0 & \cdots & \cdots & 0 \\ 0 & \gamma_2 & 0 & \cdots & 0 \\ \vdots & 0 & \ddots & \ddots & \vdots \\ \vdots & \vdots & \ddots & \ddots & 0 \\ 0 & 0 & \cdots & 0 & \gamma_N \end{bmatrix} \qquad (4.33)$$

we have $\mathbf{R}_{m_p}^{(k_p)} = \mathbf{\Phi}\mathbf{\Gamma}\mathbf{\Phi}^\top$ so that

$$\mathbf{R}_{m_p}^{(k_p)} + \phi_N\phi_N^\top = \mathbf{\Phi}\left(\mathbf{\Gamma} + \begin{bmatrix} 0 & \cdots & \cdots & 0 \\ \vdots & \ddots & \cdots & \vdots \\ \vdots & \vdots & 0 & 0 \\ 0 & \cdots & \cdots & 1 \end{bmatrix}\right)\mathbf{\Phi}^\top \qquad (4.34)$$

This allows us to rewrite equation (4.31) as

$$e^{-2\Delta_p} = \left\| \begin{bmatrix} 1 & \cdots & \cdots & 0 \\ \vdots & \ddots & \cdots & \vdots \\ \vdots & \vdots & 1 & 0 \\ 0 & \cdots & \cdots & \frac{\gamma_N}{1+\gamma_N} \end{bmatrix} + \mathbf{\Gamma}^{-1/2}\mathbf{\Phi}^\top\mathbf{s}_{m_p}^{(k_p)}\mathbf{s}_{m_p}^{(k_p)\top}\mathbf{\Phi}\mathbf{\Gamma}^{-1/2} \right\| \quad (4.35)$$

Finally, we can rewrite $\mathbf{s}_{m_p}^{(k_p)}$ as

$$\mathbf{s}_{m_p}^{(k_p)} = \sum_{i=1}^{N} a_i\phi_i \qquad (4.36)$$

where $\sum_i a_i^2 = 1$. We then have

$$e^{-2\Delta_p} = \left\| \begin{bmatrix} 1 & \cdots & \cdots & 0 \\ \vdots & \ddots & \cdots & \vdots \\ \vdots & \vdots & 1 & 0 \\ 0 & \cdots & \cdots & \frac{\gamma_N}{1+\gamma_N} \end{bmatrix} + \begin{bmatrix} \frac{a_1}{\sqrt{\gamma_1}} \\ \frac{a_2}{\sqrt{\gamma_2}} \\ \vdots \\ \frac{a_{N-1}}{\sqrt{\gamma_{N-1}}} \\ \frac{a_N}{\sqrt{1+\gamma_N}} \end{bmatrix} \begin{bmatrix} \frac{a_1}{\sqrt{\gamma_1}} & \frac{a_2}{\sqrt{\gamma_2}} & \cdots & \frac{a_{N-1}}{\sqrt{\gamma_{N-1}}} & \frac{a_N}{\sqrt{1+\gamma_N}} \end{bmatrix} \right\| \quad (4.37)$$

Now, we note that there are $N - 2$ vectors of the form

$$\psi_i = \begin{bmatrix} \psi_{i1} \\ \psi_{i2} \\ \vdots \\ \psi_{i(N-1)} \\ 0 \end{bmatrix}, \quad i = 1, 2, ..., N-2, \quad \text{for which} \quad \psi_i^\top \begin{bmatrix} \frac{a_1}{\sqrt{\gamma_1}} \\ \frac{a_2}{\sqrt{\gamma_2}} \\ \vdots \\ \frac{a_{N-1}}{\sqrt{\gamma_{N-1}}} \\ \frac{a_N}{\sqrt{1+\gamma_N}} \end{bmatrix} = 0$$

We then note that if we define

$$
\begin{aligned}
\mathbf{Z} \;=\; & \begin{bmatrix} 1 & \cdots & \cdots & & 0 \\ \vdots & \ddots & \cdots & & \vdots \\ \vdots & \vdots & 1 & & 0 \\ 0 & \cdots & \cdots & & \frac{\gamma_N}{1+\gamma_N} \end{bmatrix} \\[2mm]
+ \; & \begin{bmatrix} \frac{a_1}{\sqrt{\gamma_1}} \\ \frac{a_2}{\sqrt{\gamma_2}} \\ \vdots \\ \frac{a_{N-1}}{\sqrt{\gamma_{N-1}}} \\ \frac{a_N}{\sqrt{1+\gamma_N}} \end{bmatrix} \begin{bmatrix} \frac{a_1}{\sqrt{\gamma_1}} & \frac{a_2}{\sqrt{\gamma_2}} & \cdots & \frac{a_{N-1}}{\sqrt{\gamma_{N-1}}} & \frac{a_N}{\sqrt{1+\gamma_N}} \end{bmatrix}
\end{aligned} \tag{4.38}
$$

then $\mathbf{Z}\psi_i = \psi_i$ for $i = 1, 2, ..., N-2$ so the ψ_i are eigenvectors of $\mathbf{Z}$ and the remaining eigenvectors reside in a space spanned by the orthogonal vectors

$$\mathbf{x}_1 = \begin{bmatrix} \frac{a_1}{\sqrt{\gamma_1}} \\ \frac{a_2}{\sqrt{\gamma_2}} \\ \vdots \\ \frac{a_{N-1}}{\sqrt{\gamma_{N-1}}} \\ 0 \end{bmatrix} \quad \text{and} \quad \mathbf{x}_2 = \begin{bmatrix} 0 \\ \vdots \\ 0 \\ 1 \end{bmatrix} \tag{4.39}$$

which allows us to define the unitary matrix

$$\mathbf{U} = \begin{bmatrix} \psi_1 & \psi_2 & \cdots & \psi_{N-2} & \mathbf{x}_1/|\mathbf{x}_1| & \mathbf{x}_2 \end{bmatrix} \tag{4.40}$$

such that

$$|\mathbf{U}^\top \mathbf{Z} \mathbf{U}| = \begin{bmatrix} 1 & 0 & \cdots & & \cdots & 0 \\ \vdots & \ddots & \cdots & & \cdots & \vdots \\ \vdots & \vdots & 1 & & \cdots & 0 \\ \vdots & \vdots & 0 & |\mathbf{x}_1|^2 + 1 & & |\mathbf{x}_1|\frac{a_N}{\sqrt{1+\gamma_N}} \\ 0 & \cdots & \cdots & |\mathbf{x}_1|\frac{a_N}{\sqrt{1+\gamma_N}} & & \frac{\gamma_N + a_N^2}{1+\gamma_N} \end{bmatrix} \tag{4.41}$$

Since $|\mathbf{Z}| = |\mathbf{U}^\top \mathbf{Z} \mathbf{U}|$ we have

$$|\mathbf{Z}| = \frac{1}{1 + \gamma_N} \left(\gamma_N + \sum_{i=1}^{N} a_i^2 \frac{\gamma_N}{\gamma_i} \right) \tag{4.42}$$

We now consider the constraints on the a_i when Δ_p is small. We have

$$e^{-2\Delta_p} = 1 - \epsilon = \frac{1}{1 + \gamma_N} \left(\sum_{i=1}^{N} a_i^2 \frac{\gamma_N}{\gamma_i} + \gamma_N \right) \tag{4.43}$$

where $\epsilon > 0$. We can rewrite this as

$$1 - \epsilon(1 + \gamma_N) = 1 - \epsilon' = \sum_{i=1}^{N} a_i^2 \frac{\gamma_N}{\gamma_i} \tag{4.44}$$

which we can further rewrite as

$$\epsilon' = \sum_{i=1}^{N} a_i^2 (1 - \frac{\gamma_N}{\gamma_i}) \tag{4.45}$$

since $\sum_{i=1}^{N} a_i^2 = 1$. We then note that the coefficients $\gamma_N/\gamma_i < 1$ unless $\gamma_i = \gamma_N$ since we assumed eigenvalues ordered from largest to smallest so that all terms in the sum are non-negative. Therefore,

$$\epsilon' \geq a_i^2 (1 - \frac{\gamma_N}{\gamma_i}) \tag{4.46}$$

$i = 1, 2, ..., N$. so as $\epsilon' \to 0$ we must have each term in the sum approach zero as well. Thus, for large enough p, we cannot simultaneously have large a_i and γ_N/γ_i differing greatly from 1 – which implies that in the limit of $p \to \infty$ ($\epsilon' \to 0$), the codeword to be replaced, $s_{m_p}^{(k_p)}$ approaches a minimum eigenvector of $\mathbf{R}_{m_p}^{(k_p)}$. Finally, since it was assumed that $s_{m_p}^{(k_p)}$ was chosen to maximally increase the sum capacity, for any other codeword $s_m^{(k)}$ the corresponding change in sum capacity could only be smaller were that codeword replaced. Therefore, if at algorithm step p the a_i associated with $s_{m_p}^{(k_p)}$ are tightly bound by equation (4.46), then the corresponding coefficients for every other codeword in the system are similarly bound at step p as well. Thus, all the codewords in the ensemble simultaneously approach minimum eigenvectors of their corresponding $\mathbf{R}_m^{(k)}$. And since $\mathbf{R}^{(k)} = \mathbf{R}_m^{(k)} + s_m^{(k)} s_m^{(k)\top}$, the maximum capacity increase algorithm for interference avoidance produces codeword ensembles in the limit for which all codewords $s_m^{(k)}$ are eigenvectors of $\mathbf{R}^{(k)}$ – which proves the desired result.

Next, we note that there may exist fixed points where all codewords are eigenvectors of their associated inverted channel covariance matrices, but which do not correspond to water filling solutions [55]. However, by [54], these points are suboptimal and can be escaped as part of the algorithm and sum capacity further increased. Since the maximum sum capacity increase algorithm for interference avoidance does not stop unless each user codeword ensemble is a water filling solution to that user's inverted channel problem, its application must result in water filling codeword ensembles $\mathbf{S}_\ell$ for all users.

So in summary, there are at least two algorithms based on greedy interference avoidance that can be used for codeword optimization with dispersive channels, and which are guaranteed to converge to maximum sum capacity ensembles of codewords. We note that, while the multiuser eigen-algorithm is an instance of iterative water filling [86], the maximum sum capacity increase algorithm is not, although both converge to simultaneous water filling solutions.

3. Additional Properties of Optimal Codeword Ensembles

In this section we discuss additional properties of optimal codeword ensembles derived using greedy interference avoidance. We note that these properties have been mentioned informally for multiuser DMT systems [17] and can also be found in recent work dealing with multicarrier systems [38]. We also note that we look at these properties from a codeword perspective rather than a transmit covariance matrix perspective as in previous work [16, 17, 38].

The properties that we will refer to are straightforward consequences of the simultaneous water filling solution that is satisfied by users in their respective inverted channel problems at the optimal point, and which corresponds to the codeword ensemble that maximizes sum capacity.

We start by mentioning that for the optimal codeword ensemble obtained through application greedy interference avoidance, all symbols of any given user have the same SINR. This property, implies that a uniform receiver structure can be implemented, which could be attractive from a practical standpoint for integration purposes.

Another interesting property is that the optimal received signal covariance matrix $\mathbf{R}$ as well as all inverted channel covariance matrices $\mathbf{R}^{(\ell)}$ are diagonal. This is a consequence of the diagonal noise covariance and channel gain matrices used in our model. In order to see this, we note that the sum capacity expression

$$C = \frac{1}{2} \log \left[\det \left(\sum_{\ell=1}^{L} \mathbf{\Lambda}_\ell \mathbf{S}_\ell \mathbf{S}_\ell^\top \mathbf{\Lambda}_\ell + \mathbf{W} \right) \right] - \frac{1}{2} \log(\det \mathbf{W}) \qquad (4.47)$$

can also be rewritten as

$$C = \frac{1}{2} \log \left[\det \left(\sum_{\ell=1}^{L} \tilde{\Lambda}_\ell \mathbf{S}_\ell \mathbf{S}_\ell^\top \tilde{\Lambda}_\ell + \mathbf{I} \right) \right] \tag{4.48}$$

with

$$\tilde{\Lambda}_\ell = \mathbf{W}^{-1/2} \Lambda_\ell = \mathrm{diag} \left\{ \frac{\lambda_1^{(\ell)}}{\sigma_1}, \ldots, \frac{\lambda_n^{(\ell)}}{\sigma_n}, \ldots, \frac{\lambda_N^{(\ell)}}{\sigma_N} \right\} \qquad \ell = 1, \ldots, L \tag{4.49}$$

Equation (4.48) can also be thought of as representing sum capacity for an N-dimensional multiaccess vector channel with L users, in which each user transmits a fraction $p_n^{(\ell)}$ of its total power over scalar channel $n = 1, \ldots, N$ with corresponding gain $g_n^{(\ell)} = \lambda_n^{(\ell)}/\sigma_n$ corrupted by additive white Gaussian noise with unit variance. For such a scalar multiaccess channel sum capacity is equal to [15, p. 405]

$$C_n = \frac{1}{2} \log \left(1 + \sum_{\ell=1}^{L} g_n^{(\ell)2} p_n^{(\ell)} \right) \tag{4.50}$$

and the sum capacity of the corresponding N-dimensional multiaccess vector channel can be written as

$$C = \frac{1}{2} \sum_{n=1}^{N} C_n = \frac{1}{2} \log \prod_{n=1}^{N} \left(1 + \sum_{\ell=1}^{L} g_n^{(\ell)2} p_n^{(\ell)} \right) \tag{4.51}$$

We note that the sum capacity value in equation (4.51) is the information theoretic upper bound on sum capacity of a multiaccess vector channel and that the sum capacity value in equation (4.48) is equal to that in equation (4.51) only at the optimal point corresponding to the simultaneous water filling solution.

The fraction of power transmitted by user ℓ over scalar channel n is given by

$$p_n^{(\ell)} = \sum_{m=1}^{M_\ell} s_{mn}^{(\ell)2} \tag{4.52}$$

and is obtained by summing the squared component n of all codewords $m = 1, \ldots, M_\ell$ of user ℓ. We note that $p_n^{(\ell)}$ is the nth element of the main diagonal of $\mathbf{S}_\ell \mathbf{S}_\ell^\top$. Therefore, the term $g_n^{(\ell)2} p_n^{(\ell)}$ represents the nth diagonal element of $\tilde{\Lambda}_\ell \mathbf{S}_\ell \mathbf{S}_\ell^\top \tilde{\Lambda}_\ell$, and the term $1 + \sum_{\ell=1}^{L} g_n^{(\ell)2} p_n^{(\ell)}$ is the nth diagonal element of $\sum_{\ell=1}^{L} \tilde{\Lambda}_\ell \mathbf{S}_\ell \mathbf{S}_\ell^\top \tilde{\Lambda}_\ell + \mathbf{I}$. This implies that the determinant of the matrix that appears in the sum capacity expression in equation (4.48) is actually equal to the product of its diagonal elements. By Hadamard inequality [27, p. 477] we know that for a positive definite matrix, as is the case with $\sum_{\ell=1}^{L} \tilde{\Lambda}_\ell \mathbf{S}_\ell \mathbf{S}_\ell^\top \tilde{\Lambda}_\ell + \mathbf{I}$, the determinant is equal to the product of its diagonal elements if and only if the matrix is diagonal. As a consequence, this implies the desired result, namely that the optimal received signal covariance matrix $\mathbf{R}$, as well as all inverted channel covariance matrices $\mathbf{R}^{(\ell)}$ are diagonal.

The following properties deal with signal space partition among users at the optimal point, and potential overlap between subspaces in which users reside. We first note that if codeword matrices of two given users each spans the whole signal space, then these users must have identical channels. To illustrate, consider equations (4.12) and (4.15) for any pair of distinct users i and j at the optimal point

$$\begin{aligned}
\mathbf{\Lambda}_i^{-1}\mathbf{R}\mathbf{\Lambda}_i^{-1}\mathbf{S}_i\mathbf{S}_i^\top &= \mu_i\mathbf{S}_i\mathbf{S}_i^\top \\
\mathbf{\Lambda}_j^{-1}\mathbf{R}\mathbf{\Lambda}_j^{-1}\mathbf{S}_j\mathbf{S}_j^\top &= \mu_j\mathbf{S}_j\mathbf{S}_j^\top
\end{aligned} \tag{4.53}$$

If the two distinct users i and j both span the signal space then matrices $\mathbf{S}_i\mathbf{S}_i^\top$ and $\mathbf{S}_j\mathbf{S}_j^\top$ are invertible, and by post-multiplying with their inverses followed by appropriate multiplication by the corresponding channel matrix we get

$$\mathbf{R} = \mu_i\mathbf{\Lambda}_i = \mu_j\mathbf{\Lambda}_j \tag{4.54}$$

This implies that the channel matrix corresponding to user i is a scaled version of user j channel matrix

$$\mathbf{\Lambda}_i = \frac{\mu_j}{\mu_i}\mathbf{\Lambda}_j \tag{4.55}$$

or in other words, that users i and j see the same channel. This remark is an indication that the signal space is frequency partitioned at the optimal point. That is, codeword matrices of users with different channels cannot contain all frequency components.

Now suppose we assume distinct channels for each user with the addition proviso that *the ratio of channel gain magnitudes for any pair of users $i \neq j$ is different for different dimensions $r \neq s$ corresponding to different frequencies $f_r \neq f_s$*, that is

$$\frac{|\lambda_r^{(i)}|}{|\lambda_r^{(j)}|} \neq \frac{|\lambda_s^{(i)}|}{|\lambda_s^{(j)}|} \quad \forall i \neq j \in \{1,\dots,L\}, \ r \neq s, \ f_r \neq f_s \tag{4.56}$$

This is a reasonable assumption for some level of precision ε in the representation of channel eigenvalue matrices since a small perturbation $O(\varepsilon)$ will spoil any potential equality.

Under this assumption we note that at the optimal point no two users can reside in subspaces that overlap in more than one frequency. This property has been observed in the context of discrete multitone systems by S. Diggavi [16, 17]. To see this we use again equation (4.12) to write for any distinct users i and j

$$\mathbf{R} = \mathbf{\Lambda}_i\mathbf{R}^{(i)}\mathbf{\Lambda}_i = \mathbf{\Lambda}_j\mathbf{R}^{(j)}\mathbf{\Lambda}_j \tag{4.57}$$

and note that the diagonal structure of matrices in equation (4.57) implies that for any dimension r in which users i and j overlap we have

$$\mu_i |\lambda_r^{(i)}|^2 = \mu_j |\lambda_r^{(j)}|^2 \tag{4.58}$$

This can also be rewritten as

$$\frac{|\lambda_r^{(i)}|^2}{|\lambda_r^{(j)}|^2} = \frac{\mu_j}{\mu_i} \tag{4.59}$$

According to equation (4.56), the ratio of channel gains for users i and j is differs for distinct dimensions corresponding to different frequencies. Thus equation (4.59) is true for one and only one frequency f_r, which implies that any pair of users with codeword matrices $\mathbf{S}_i \neq \mathbf{S}_j$ can overlap at most in one frequency at the optimal point.

Finally we note that if two users i and j reside in overlapping subspaces, then the overlap occurs in the minimum gain ratio dimension over all dimensions occupied by the user. This property has been noted in recent work on multicarrier modulation [38]. In order to see this we note that the simultaneous water filling solution implies that for any user i the eigenvalue μ_i in equation (4.15) is the minimum eigenvalue of $\mathbf{R}^{(i)}$. Thus, for any pair of users i and j we can write

$$\mu_j \leq \frac{|\lambda_s^{(i)}|^2}{|\lambda_s^{(j)}|^2} \mu_i \quad \forall\, s = 1, \ldots, N \tag{4.60}$$

If users i and j overlap in dimension r then

$$\mu_j = \frac{|\lambda_r^{(i)}|^2}{|\lambda_r^{(j)}|^2} \mu_i \leq \frac{|\lambda_s^{(i)}|^2}{|\lambda_s^{(j)}|^2} \mu_i \tag{4.61}$$

Equation (4.61) indicates that overlap occurs in a dimension r which corresponds to a *minimum* gain ratio for user i over user j. Since equation (4.61) can be rewritten as

$$\mu_i = \frac{|\lambda_r^{(j)}|^2}{|\lambda_r^{(i)}|^2} \mu_j \leq \frac{|\lambda_s^{(j)}|^2}{|\lambda_s^{(i)}|^2} \mu_j \tag{4.62}$$

the same is true for user j, i.e. dimension r corresponds to a *minimum* gain ratio for user j over user i as well. Note that the minimum gain ratio for a given user is taken only over those dimensions that are actually occupied by that user.

Finally, it is also worth pointing out that the property that in the optimal codeword ensemble two users cannot overlap in more than one frequency generalizes (for a fixed number of users) as the number of frequencies that span the signal space $N \to \infty$ to distinct frequency bands for different users[1]. Such

---

[1]As the number of frequencies that span the signal space increases to infinity, overlap is on a set of zero measure, which means essentially that different users do not overlap at all.

Frequency Division Multiple Access (FDMA) is well-known to maximize the sum capacity of multiple access channels with ISI [10].

4.   Numerical Examples

We provide numerical examples that illustrate the ensemble properties we have discussed analytically. We start with a simple example of $L = 2$ users in a signal space spanned by $N = 3$ frequencies. The channel gains have been generated randomly from a uniform $[0, 1]$ distribution, and in the real notation used in the paper are given by the diagonal matrices

$$\Lambda_1 = \text{diag}\{0.9501, 0.9501, 0.2311, 0.2311, 0.6068, 0.6068\}$$

$$\Lambda_2 = \text{diag}\{0.4860, 0.4860, 0.8913, 0.8913, 0.8913, 0.8913\}$$

Background noise is white with covariance matrix $\mathbf{W} = 0.1\mathbf{I}_6$. Initial user codeword matrices have also been generated randomly, and after interference avoidance is performed we get the optimal codeword matrices

$$\mathbf{S}_1 = \begin{bmatrix} -0.6839 & -0.9287 & 0.2542 & -0.5889 & -0.5243 & -0.8021 \\ -0.7264 & -0.3391 & 0.9625 & 0.6574 & 0.7282 & 0.3584 \\ 0 & 0 & 0 & 0 & 0 & 0 \\ 0 & 0 & 0 & 0 & 0 & 0 \\ -0.0561 & -0.0298 & 0.0672 & 0.1011 & -0.3900 & 0.2844 \\ -0.0378 & -0.1473 & -0.0661 & -0.4591 & 0.2067 & 0.3837 \end{bmatrix}$$

$$\mathbf{S}_2 = \begin{bmatrix} 0 & 0 & 0 & 0 & 0 & 0 \\ 0 & 0 & 0 & 0 & 0 & 0 \\ -0.8332 & 0.0987 & -0.1800 & 0.7432 & 0.4777 & -0.3377 \\ -0.3767 & -0.1830 & 0.2566 & 0.5001 & -0.8048 & 0.7013 \\ 0.2332 & 0.9306 & 0.3276 & 0.2180 & 0.3406 & 0.4836 \\ 0.3309 & -0.3014 & 0.8913 & 0.3874 & 0.0896 & -0.4004 \end{bmatrix}$$

Obviously the codeword matrices do not span the whole signal space. More precisely, in this case user 1 spans the subspace determined by frequencies 1 and 3, while user 2 spans the subspace corresponding to frequencies 2 and 3. The water filling distribution that corresponds to each user's inverted channel problem can be observed by looking at the covariance matrices which are

$$\mathbf{R}^{(1)} = \text{diag}\{2.7707, 2.7707, 26.1235, 26.1235, 2.7707, 2.7707\}$$

$$\mathbf{R}^{(2)} = \text{diag}\{10.5906, 10.5906, 1.7568, 1.7568, 1.7568, 1.7568\}$$

One can also see that users overlap in frequency 3 which corresponds to minimum gain ratio for both users.

We provide another example, this time using the complex framework, and with more users and spanning frequencies than before, so that the overlap properties may be better observed. Specifically, this time we consider $L = 3$ users in a signal space spanned by $N = 6$ frequencies. Channel gains are now complex and have been generated randomly with magnitudes from a uniform $[0, 1]$ distribution and phases from a uniform $[0, 2\pi]$ distribution, and are given by

$$\Lambda_1 = \text{diag}\{ \quad 0.6846 + 0.1147j, \ -1.2257 - 0.6181j, \ -1.1892 + 0.9122j$$
$$-1.0797 - 0.3725j, \ -1.2828 - 0.5019j, \ -1.1035 + 0.2680j \ \}$$

$$\Lambda_2 = \text{diag}\{ \quad -0.9718 - 0.5980j, \; 0.6201 - 0.1887j, \; -0.0201 + 1.2044j$$
$$-1.1099 + 0.4045j, \; 0.9421 - 0.9676j, \; -0.5317 + 0.1375j \; \}$$

$$\Lambda_3 = \text{diag}\{ \quad 1.1239 + 0.9118j, \; -0.1903 - 0.8703j, \; -0.2968 + 1.2641j$$
$$0.8515 + 0.8314j, \; 0.1989 + 0.5522j, \; -0.6658 + 0.0941j \; \}$$

Background noise is also white with covariance matrix $\mathbf{W} = 0.1\mathbf{I}_6$. Initial user codeword matrices have also been generated randomly, and after interference avoidance is performed we get the optimal codeword matrices

$$\mathbf{S}_1 = \begin{bmatrix} 0 & 0 & 0 \\ 0.5577 - 0.2000j & -0.6635 + 0.1138j & 0.6596 - 0.1263j \\ 0.2376 + 0.1994j & -0.1306 + 0.2040j & 0.0585 + 0.2054j \\ 0 & 0 & 0 \\ 0 & 0 & 0 \\ 0.7435 & 0.6986 & 0.7095 \end{bmatrix}$$

$$\begin{bmatrix} 0 & 0 & 0 \\ -0.2676 + 0.4501j & 0.6358 & 0.7429 \\ -0.3773 - 0.3602j & -0.5437 - 0.1581j & 0.2580 - 0.4668j \\ 0 & 0 & 0 \\ 0 & 0 & 0 \\ 0.6736 & -0.4682 - 0.2364j & 0.0793 + 0.3967j \end{bmatrix}$$

$$\mathbf{S}_2 = \begin{bmatrix} 0 & 0 & 0 \\ 0 & 0 & 0 \\ 0 & 0 & 0 \\ 0.7710 & 0.1903 + 0.2344j & -0.1810 - 0.4009i \\ 0.1965 - 0.6057j & 0.9533 & 0.8981 \\ 0 & 0 & 0 \end{bmatrix}$$

$$\begin{bmatrix} 0 & 0 & 0 \\ 0 & 0 & 0 \\ 0 & 0 & 0 \\ 0.3776 - 0.1337j & 0.9412 & 0.8247 \\ 0.9163 & -0.2769 - 0.1934j & -0.3100 + 0.4731j \\ 0 & 0 & 0 \end{bmatrix}$$

$$\mathbf{S}_3 = \begin{bmatrix} 0.9752 & 0.9460 & 0.7399 \\ 0 & 0 & 0 \\ 0.0465 - 0.2134j & -0.2049 + 0.0750j & 0.6422 + 0.0888j \\ 0.0103 + 0.0351j & 0.1069 - 0.2146j & 0.0334 - 0.1768j \\ 0 & 0 & 0 \\ 0 & 0 & 0 \end{bmatrix}$$

$$\begin{bmatrix} -0.2545 + 0.1283j & -0.3400 - 0.5552j & 0.6774 \\ 0 & 0 & 0 \\ 0.8437 & 0.7240 & 0.1983 - 0.3280j \\ 0.3826 - 0.2459j & -0.1305 + 0.1869j & 0.0775 + 0.6231j \\ 0 & 0 & 0 \\ 0 & 0 & 0 \end{bmatrix}$$

Again, the optimal codeword matrices do not span the whole signal space. In this case user 1 spans the subspace determined by frequencies 2, 3, and 6; user 2 spans the subspace corresponding to frequencies 4 and 5; while user 3 spans the subspace implied by frequencies 1, 3, and 4. The water filling distribution that corresponds to each user's inverted channel problem can be observed by looking at the covariance matrices which are

$$\mathbf{R}^{(1)} = \text{diag}\{15.0089, 2.5918, 2.5918, 3.7487, 3.3680, 2.5918\}$$

$$\mathbf{R}^{(2)} = \text{diag}\{5.5547, 11.6265, 4.0123, 3.5041, 3.5041, 11.0805\}$$

$$\mathbf{R}^{(3)} = \text{diag}\{3.4527, 6.1543, 3.4527, 3.4527, 18.5505, , 7.3916\}$$

Overlap between users is also observed as follows: users 1 and 3 overlap in frequency 3, and users 2 and 3 overlap in frequency 4; both overlaps occur in minimum gain ratio dimensions.

5.   Fading Channels

Our treatment of frequency dependent channel gains has set the stage for time-varying frequency dependence as occurs over multipath wireless channels. As time variations appear to be unpredictable, wireless channels have randomly time varying impulse responses and are characterized statistically – which further implies that the sum capacity will also be a random variable. In this section we show how application of interference avoidance can be extended from dispersive channels, which are deterministic and time-invariant, to fading channels, which are time-varying with random realizations. We concentrate our attention on indoor wireless channels for which extensive studies have been performed (see [22] and references therein), and assume a frequency selective fading channel model with flat Rayleigh fading of the carriers. This model agrees with our system description in equation (4.5), and has been used previously for indoor wireless channels [84]. In a Rayleigh fading environment the amplitude scaling $\kappa_n^{(\ell)}$ of the n-th carrier due to the channel between user ℓ and base station is a Rayleigh random variable with the probability density function

$$f_{\kappa_n^{(\ell)}}(\kappa_n^{(\ell)}) = \frac{\kappa_n^{(\ell)}}{\sigma_n^{(\ell)2}} e^{-\frac{\kappa_n^{(\ell)2}}{2\sigma_n^{(\ell)2}}} \tag{4.63}$$

where the parameter $\sigma_n^{(\ell)2}$ is related to the second moment of the Rayleigh random variable $E[\kappa_n^{(\ell)2}] = 2\sigma_n^{(\ell)2}$. The second moment of this random variable characterizes the average channel and represents the gain corresponding to carrier n of user ℓ average channel, i.e. $E[\kappa_n^{(\ell)2}] = 2\sigma_n^{(\ell)2} = \lambda_n^{(\ell)}$.

5.1   Slowly Fading Channels

A slowly fading channel has a large coherence time T_c, or equivalently a small Doppler spread [48]. This is a measure of the channel's temporal variation and for residential buildings or office environments with reduced mobility one can assume that a channel is stable for a few seconds [22]. For high data rates this implies that the channel is stable for a large number of symbols thus allowing identification of channel parameters. In our case these are the amplitude scalings of different carriers (or channel eigenvalues) and can be determined by probing the channel with appropriate pilot symbols [37].

To apply interference avoidance algorithms, an estimate of the received signal autocorrelation matrix must be obtained in addition to channel identification. By subtracting the contribution of a given symbol/user the autocorrelation matrix of the corresponding interference-plus-noise process needed by the minimum eigenvector algorithm is then computed. This problem has been studied extensively in [67] where it has been shown that the minimum number of symbol intervals needed for accurate estimation of the autocorrelation matrix of the received signal depends on dimensionality (number of carriers in our case) and on the number of codewords/users in the system. The use of training sequences to speed up estimation is also presented in [67].

In order for interference avoidance to be efficient, the overhead associated with identification, estimation, and application of the interference avoidance algorithm, should be small compared to the actual number of information symbols sent. A symbol duration $T = 2\mu s$ (which corresponds to 0.5 Msymbols/s and also satisfies the condition imposed by the multicarrier modulation scheme[2]) implies that approximately 1,000,000 symbols can be transmitted in 2 s, which is of the order of the channel coherence time. With a 10% overhead factor this implies that up to 100,000 symbols may be used for estimation, which seems to cover a lot of potential combinations of dimensions/users/codewords [67].

In conclusion, for slowly fading indoor channels, it may be possible to apply interference avoidance in a straightforward way. After channel eigenvalues are determined, interference avoidance algorithms are applied to compute optimal codeword ensembles that maximize sum capacity. These codeword ensembles are then used for transmission, with periodic updates coupled to channel variation.

---

[2]T must be larger than the duration of all channel impulse responses. In the case of multipath fading channels this is equivalent to the duration of the multipath intensity profile which for indoor environments (medium size office or residential buildings) is of the order of 400 ns [22].

5.2 Fast Fading Channels and Interference Avoidance for the Average Channel

Now suppose the channel is dynamic and cannot be estimated rapidly enough to satisfy the quasistatic conditions necessary to apply interference avoidance for each channel instance. That is, suppose that by the time the channel is estimated and the codewords calculated, the channel has already changed to a new realization.

In this case only average characteristics of the channel can be measured, and we seek to determine the performance of applying interference avoidance techniques to the average channel. For the multicarrier modulation scheme proposed, with a frequency selective fading channel model, we use the average values of the gains in equation (4.63) to perform interference avoidance and determine optimal codeword ensembles for the average channel. Let us denote the ensemble of user codeword matrices by $\mathcal{S} = \{\mathbf{S}_1, \ldots, \mathbf{S}_L\}$ and the ensemble of channel eigenvalue matrices by $\Lambda = \{\Lambda_1, \ldots, \Lambda_L\}$. Then, for a given ensemble of codeword and channel gain matrices the sum capacity obtained from equation (2.14) is

$$C_S(\Lambda, \mathcal{S}) = \frac{1}{2} \log \left[\det \left(\sum_{\ell=1}^{L} \Lambda_\ell \mathbf{S}_\ell \mathbf{S}_\ell^\top \Lambda_\ell + \mathbf{W} \right) \right] - \frac{1}{2} \log \left(\det \mathbf{W} \right) \quad (4.64)$$

For a fixed ensemble of codeword matrices $\mathcal{S}$, due to the randomness of channel realizations Λ, the resulting sum capacity is a random variable. Using Jensen's inequality [15, p. 25] and the fact that the function $\log(\det \mathbf{A})$ is concave for positive semidefinite matrices $\mathbf{A}$ [27, p. 466] we can write that

$$E_\Lambda[C_S(\Lambda, \mathcal{S})] \leq C_S(\bar{\Lambda}, \mathcal{S}) \quad (4.65)$$

where $\bar{\Lambda}$ denotes the average value of the channel eigenvalue matrices ensemble. Moreover, for any ensemble of user codeword matrices $\mathcal{S}$ we can also write that

$$C_S(\bar{\Lambda}, \mathcal{S}) \leq C_S(\bar{\Lambda}, \mathcal{S}^*) \quad (4.66)$$

where

$$\mathcal{S}^* = \arg \max_{\mathcal{S}} C_S(\bar{\Lambda}, \mathcal{S}) \quad (4.67)$$

is the ensemble of codeword matrices that maximizes the sum capacity corresponding to the ensemble of average channel gain matrices $\bar{\Lambda}$, and is obtained by applying interference avoidance algorithms with average values of channel parameters.

We note that the codeword ensemble which is optimal for the average channel is

$$\mathcal{S}^{\dagger} = \arg \max_{\mathcal{S}} E_{\boldsymbol{\Lambda}}[C_{\mathsf{s}}(\boldsymbol{\Lambda}, \mathcal{S})] \tag{4.68}$$

and that, although $\mathcal{S}^*$ is not actually the codeword ensemble optimal for the average channel, empirically we have seen improvements when using $\mathcal{S}^*$. The improvements can be observed by looking at sum capacity CCDFs, which show what capacity can be achieved with a given probability for a given ensemble of codewords. An outage occurs whenever capacity is below a given value, and the probability of outage P_{out} can be identified from the corresponding CCDF.

The sum capacity CCDFs presented in figures 4.4 and 4.5, which are typical for the experiments we have performed, correspond to a system using $N = 10$ carriers, with $L = 2$ users, and white noise at the base station receiver with a signal-to-noise ratio of 10 dB. A set of 1,000 randomly chosen codeword ensembles have been generated and 10,000 realizations of Rayleigh fading channels corresponding to the two users have been considered. Codeword ensembles optimal for the average channel and codeword ensembles optimal for each channel realization were also computed. In figure 4.4 the average channel was considered ideal, with all channel eigenvalues equal to 1, which implies that second moments of Rayleigh random variables $\kappa_n^{(\ell)^2}$ in equation (4.63) are also equal to 1. In the case of a non-ideal average channel, channel eigenvalues can be obtained from the frequency spectrum corresponding to the average channel. For the CCDFs presented in figure 4.5 a very general case of non-ideal average channel was considered, with channels eigenvalues uniformly distributed between 0.5 and 1.5 corresponding to an average channel that attenuates some frequencies and boosts others.

From both figures (4.4 and 4.5) we note that for the same probability of outage the sum capacity when using codeword ensembles yielded by application of interference avoidance for the average channel is always larger than the corresponding value for random codeword ensembles. Of course, the best performance is obtained with codeword ensembles optimal for each channel realization, but under rapid channel variation this may not always be possible. Thus, our simulation results confirm that using codeword ensembles tuned to the average channel is better than using random codewords but worse than using per-channel codewords. However, we also note that the capacity gain obtained from optimal codeword ensembles for each channel realization as compared to that obtained using average channel optimal codewords might not justify the computational burden of doing interference avoidance for each channel realization. Thus, selecting codeword sets optimal for the average channel may be a reasonable compromise.

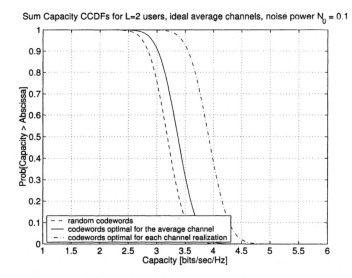

Figure 4.4. Sum Capacity CCDFs for multiaccess fading channels comparing random code-word ensembles with codeword ensembles optimal for the average channel and codeword ensembles optimal for each channel realization. Average channels are assumed to be ideal.

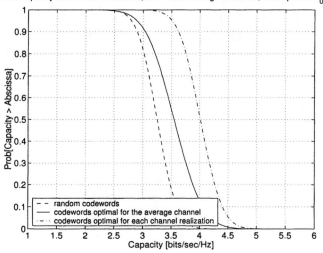

Figure 4.5. Sum Capacity CCDFs for multiaccess fading channels comparing random code-word ensembles with codeword ensembles optimal for the average channel and codeword ensembles optimal for each channel realization. Average channels are non-ideal.

6.   Discrete-Time Fading Channel Models

Although we have used a model that is related to multicarrier modulation to present application of greedy interference avoidance to dispersive/fading channels, interference avoidance methods are general. To show generality, we provide an explicit formulation for interference avoidance using discrete time channel models in this section.

We consider the discrete-time multiple access channel model characterizing the uplink of a system with L users communicating with a common receiver (base station)

$$r(n) = \sum_{\ell=1}^{L} \sqrt{h_\ell(n)} x_\ell(n) + w(n) \tag{4.69}$$

where $x_\ell(n)$ and $h_\ell(n)$ are the transmitted signal and the fading process of the ℓth user respectively, and $w(n)$ is additive Gaussian noise that corrupts the received signal. This is a basic frequency non-selective multi-access fading channel model also used in [19, 68]. By stacking values of the received signal $r(n)$ in equation (4.69) for $n = 0, \dots, N - 1$ corresponding to an observation interval, we obtain the vector-matrix equation

$$\mathbf{r} = \sum_{\ell=1}^{L} \mathbf{H}_\ell^{1/2} \mathbf{x}_\ell + \mathbf{w} \tag{4.70}$$

with $\mathbf{r} = [r(0) \dots r(N-1)]^\mathsf{T}$, $\mathbf{H} = \text{diag}\{h_\ell(0), \dots, h_\ell(N-1)\}$, $\mathbf{x}_\ell = [x_\ell(0) \dots x_\ell(N-1)]^\mathsf{T}$, and $\mathbf{w} = [w(0) \dots w(N-1)]^\mathsf{T}$. Equation (4.70) can be further written as

$$\mathbf{r} = \sum_{\ell=1}^{L} \mathbf{H}_\ell^{1/2} \mathbf{S}_\ell \mathbf{b}_\ell + \mathbf{w} \tag{4.71}$$

by assuming that transmitted waveforms are obtained from

$$\mathbf{x}_\ell = \mathbf{S}_\ell \mathbf{b}_\ell \qquad \ell = 1, \dots, L \tag{4.72}$$

where $\mathbf{b}_\ell = [b_1^{(\ell)} \dots b_{M_\ell}^{(\ell)}]^\mathsf{T}$ is the symbol sequence to be transmitted by user ℓ through a codeword matrix $\mathbf{S}_\ell$. Note that equation (4.72) can be regarded as a spreading scheme in which each symbol in $\mathbf{b}_\ell$ is "spread" over N "time chips".

Equation (4.71) is identical in form with the multiaccess dispersive channels equation (4.5). There are however two main differences:

- The spreading scheme in equation (4.72) implies a time representation of waveforms as superposition of "time chips" specified by the corresponding codewords as opposed to the dispersive channels case where waveforms were represented in frequency as a superposition of sinusoids.

- Channels are described by random matrices containing values of the fading processes (channel states) as opposed to the dispersive channels case where they were described by channel gains.

Nevertheless, by working with average values for $\{\mathbf{H}_\ell\}$ matrices one can directly apply the eigen-algorithm for interference avoidance and derive an ensemble of optimal codeword matrices $\{\mathbf{S}_\ell\}$ for the average channel.

It is also worth noting here that the signal space partitioning induced by the optimal codeword ensemble leads to a time division multiple access (TDMA) scheme, in which users transmit information in distinct time slots with possible overlaps in at most one slot, using time slots with best channel states. Similar conclusions with regard to codeword ensembles apply for this case as well.

7. Summary

We have extended application of interference avoidance to channels where gain varies over dimensions for different users. The models considered apply to discrete tone (OFDM) channels as well as discrete time (sampled) channels. In all cases, an optimal codeword ensemble is reached using interference avoidance and its variants. We also considered the case of time-varying (fading) channels and found that when the channel coherence time permits a sufficient number of symbols for training without exceeding some allowable overhead limit, interference avoidance could be used on a per-channel-instance basis. Indoor channels seem to easily meet these requirements for coherence times. When the channel coherence time is too small to allow per-channel-instance interference avoidance, we have found that doing interference avoidance on the average channel offers significant advantages over random codewords.

We have not considered asynchronous operation, multiple receivers and antennas (for users and/or the base) among other topics. However, we will find that interference avoidance can be readily applied to a variety of different scenarios upon introduction of a generalized notion of interference avoidance over vector spaces – the subject of the next chapter.

Chapter 5

GENERALIZATION: INTERFERENCE AVOIDANCE
FOR MULTIACCESS VECTOR CHANNELS

Vector channel models provide a theoretical framework for the analysis of a wide variety of communication channels – multiple access channels, channels with memory, or channels with multiple antennas in the transmitter and/or receiver – and have thus of late received increasing attention from the research community. Capacity results for multiaccess vector channels have been derived in [81], and an asymptotically optimal water filling algorithm for multiaccess vector channels can be found in [80]. A characterization of the capacity region for multiaccess vector channels and an iterative water filling algorithm to evaluate optimal transmit spectra that maximize the sum capacity of the channel can be found in [86].

Code Division Multiple Access (CDMA) schemes have a natural vector channel representation implied by the signature sequences (codewords) corresponding to distinct users in the system. The selection of optimal signature sequences (or codeword ensembles) that maximize the sum capacity in a CDMA system has been addressed by several researchers and algorithms that yield such optimal codeword ensembles can be found in [56, 78]. Optimal signature sequences provide all users in the system with a uniform signal-to-interference plus noise ratio. Furthermore, it has been shown that the optimal linear receiver for such ensembles is a *matched filter* for each codeword [78]. Such optimal codeword ensembles for CDMA systems can also be obtained through application of interference avoidance methods [54, 55].

However, interference avoidance provides a *distributed* algorithm for codeword optimization in which users independently adjust codewords in response to changing patterns of interference. As opposed to centralized optimization methods performed at the receiver and based on complete knowledge of the system, distributed optimization through interference avoidance requires only that each user know its associated channel and have access to the system covariance information through a feedback channel broadcast from the base. In this case, the base could track changes in codewords similar to an adaptive equalizer. Of course, the mathematics of interference avoidance allows for centralized processing as well, but it is the distributed version of the algorithm which may prove most useful in unlicensed/uncoordinated environments.

In this chapter we extend application of greedy interference avoidance methods based on a minimum eigenvector approach to general multiaccess vector channels. As we have already seen, application of greedy interference avoidance methods yields codeword ensembles that maximize sum capacity for particular vector channel models corresponding to multiaccess dispersive channels.

We use a general multiaccess vector channel model similar to that used by Yu et al. in [86] along with a multicode CDMA scheme for transmission of information, and show how greedy interference avoidance applies for a given user through projection of the received signal onto its corresponding signal space. We then show that again, application of greedy interference avoidance for any codeword/user monotonically increases sum capacity and discuss different codeword update sequences – all of which converge to a fixed point corresponding to the simultaneously water filled user covariance ensemble which maximizes sum capacity as shown in [86]. But we must emphasize that although codeword update sequences can be designed to correspond to the iterative water filling algorithm presented in [86], application of greedy interference avoidance to codeword optimization in a multiuser system *is not in general a water filling procedure*. Rather, convergence to a simultaneously water filled solution is an *emergent property* of the algorithm.

This chapter firmly establishes how greedy interference avoidance can be applied to all communication problems in which the underlying model is a multiaccess Gaussian vector channel. In addition, the potential for distributed adaptive implementation of greedy interference avoidance in multiuser systems along with the identical receiver structures implied by a common SINR for all codewords owed to (optimal) matched filter detection, could make interference avoidance algorithms good candidates for integrated receiver structures.

1.   The Vector Multiple Access Channel

A single user vector channel is defined by [86]

$$y = Hx + n \qquad (5.1)$$

where x and y are the input and output vectors of respective dimensions N_x and N_y, n is the additive noise vector, and H is the $N_y \times N_x$ channel matrix. The vector channel in equation (5.1) represents a linear transformation from an input signal space of dimension N_x to an output signal space of dimension N_y defined by the channel matrix H. For memoryless channels the matrix H merely relates the bases of the input and output signal spaces [66, p. 116], while for

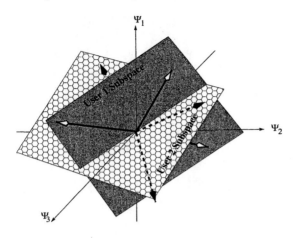

Figure 5.1. 3-dimensional receiver signal space with 2 users residing in 2-dimensional subspaces. Vectors represent particular signals in the user 1 and user 2 signal spaces (continuous and dashed lines, respectively).

channels with memory **H** also incorporates channel attenuation and multipath. We assume that **H** is known and stable for a large number of transmission frames or that its average behavior can be measured. As it was argued in the previous chapter these are not particularly onerous constraints for high data rates and office environments with reduced degrees of mobility.

Extending the definition in equation (5.1) to multiple users, a multiaccess vector channel is obtained in which a set of L users communicate with a common receiver (basestation). The multiaccess vector channel is defined by the equation

$$\mathbf{r} = \sum_{\ell=1}^{L} \mathbf{H}_\ell \mathbf{x}_\ell + \mathbf{n} \tag{5.2}$$

where $\mathbf{x}_\ell$ of dimension N_ℓ is the input vector corresponding to user ℓ, $\ell = 1, \ldots, L$, $\mathbf{r}$ of dimension N is the received vector at the common receiver corrupted by additive noise vector $\mathbf{n}$ of the same dimension, and $\mathbf{H}_\ell$ is the $N \times N_\ell$ channel matrix corresponding to user ℓ. It is assumed that $N \geq N_\ell, \forall \ell = 1, \ldots, L$. This is a general approach to a multiuser communication system in which different users reside in different signal spaces, with different dimensions and potential overlap between them, and all being subspaces of the receiver signal space. We note that each user's signal space as well as the receiver signal space are of finite dimension and are implied by a finite time interval $\mathcal{T}$ and finite bandwidths W_ℓ for each user ℓ, respectively and W (which includes all W_ℓ's corresponding to all users) for the receiver [31]. Figure 5.1 provides a graphical illustration of such a signal space configuration.

In this signal space setting we assume that in an interval of duration T each user ℓ sends a "frame" of data using a multicode CDMA approach similar to that depicted in Figure 4.1 which implies that the sequence of information symbols $\mathbf{b}_\ell = [b_1^{(\ell)} \dots b_{M_\ell}^{(\ell)}]^\top$ is transmitted as a linear superposition of distinct, unit-energy codewords $\mathbf{s}_m^{(\ell)}$

$$\mathbf{x}_\ell = \sum_{m=1}^{M_\ell} b_m^{(\ell)} \mathbf{s}_m^{(\ell)} = \mathbf{S}_\ell \mathbf{b}_\ell \tag{5.3}$$

where

$$\mathbf{S}_\ell = \begin{bmatrix} | & | & & | \\ \mathbf{s}_1^{(\ell)} & \mathbf{s}_2^{(\ell)} & \cdots & \mathbf{s}_{M_\ell}^{(\ell)} \\ | & | & & | \end{bmatrix} \tag{5.4}$$

is the $N_\ell \times M_\ell$ user ℓ codeword matrix. Therefore, with the multicode CDMA transmission of information the multiaccess vector channel equation (5.2) can be rewritten as

$$\mathbf{r} = \sum_{\ell=1}^{L} \mathbf{H}_\ell \mathbf{S}_\ell \mathbf{b}_\ell + \mathbf{n} \tag{5.5}$$

We note that if $M_\ell \geq N_\ell$ then the $N_\ell \times N_\ell$ covariance matrix of $\mathbf{x}_\ell$ defined as $\mathbf{X}_\ell = E[\mathbf{x}_\ell \mathbf{x}_\ell^\top] = \mathbf{S}_\ell \mathbf{S}_\ell^\top$ could have full rank and thereby span user ℓ signal space.

There are various ways in which one can obtain equation (5.5) for a given system. We present a general procedure of deriving it starting from an arbitrary waveform representation, and illustrate this procedure for a direct sequence CDMA system with multipath channels.

Let us consider a system with L users communicating with a common receiver over distinct, dispersive channels, characterized by impulse responses $h_\ell(t)$, $\ell = 1, \dots, L$, as is the case in the uplink of a wireless system. We assume that each user's signal space of dimension N_ℓ is spanned by the vector of functions $\boldsymbol{\Psi}^{(\ell)}(t) = [\Psi_1^{(\ell)}(t) \dots \Psi_{N_\ell}^{(\ell)}(t)]^\top$ and that the receiver signal space of dimension N is spanned by the vector of functions $\boldsymbol{\Psi} = [\Psi_1(t) \dots \Psi_N(t)]^\top$. Let $\mathcal{H}_\ell$ be the $N \times N_\ell$ matrix with orthonormal columns relating the bases of the two signal spaces, that is

$$\boldsymbol{\Psi}^{(\ell)}(t) = \mathcal{H}_\ell^\top \boldsymbol{\Psi}(t) \tag{5.6}$$

Each user in the system transmits the multicode CDMA signal

$$x_\ell(t) = \sum_{m=1}^{M_\ell} b_m^{(\ell)} s_m^{(\ell)}(t) \tag{5.7}$$

of duration $\mathcal{T}$ corresponding to a frame containing M_ℓ symbols, and composed of a superposition of waveforms scaled by the symbols. In the N_ℓ-dimensional signal space corresponding to user ℓ, each waveform can be represented in terms of the basis functions as

$$x_\ell(t) = \mathbf{\Psi}^{(\ell)}(t)^\top \mathbf{x}_\ell = \mathbf{\Psi}^{(\ell)}(t)^\top \mathbf{S}_\ell \mathbf{b}_\ell \tag{5.8}$$

where $\mathbf{x}_\ell$ is the equivalent signal vector corresponding to user ℓ which is equal to a linear superposition of unit norm codeword column vectors $\mathbf{s}_m^{(\ell)}$ which make up user ℓ codeword matrix $\mathbf{S}_\ell$.

At the receiver, the signal sent by user ℓ is $y_\ell(t)$, the convolution of $x_\ell(t)$ with user ℓ channel impulse response $h_\ell(t)$

$$y_\ell(t) = \int_0^T h_\ell(\tau) x_\ell(t - \tau) d\tau \tag{5.9}$$

Projected onto the basis functions of the receiver signal space the vector $\mathbf{y}_\ell$ is obtained such that component n of $\mathbf{y}_\ell$ is

$$
\begin{aligned}
y_n^{(\ell)} &= \int_0^T y_\ell(t) \Psi_n(t) dt \\
&= \int_0^T \int_0^T h_\ell(\tau) x_\ell(t - \tau) \Psi_n(t) d\tau dt \\
&= \int_0^T \int_0^T h_\ell(\tau) \mathbf{\Psi}^{(\ell)}(t - \tau)^\top \mathbf{x}_\ell \Psi_n(t) d\tau dt
\end{aligned}
\tag{5.10}
$$

Thus, vector $\mathbf{y}_\ell$ can be written as

$$\mathbf{y}_\ell = \mathbf{F}_\ell \mathcal{H}_\ell \mathbf{x}_\ell \tag{5.11}$$

where the $N \times N$ matrix $\mathbf{F}_\ell$ is obtained from the following equation

$$\mathbf{F}_\ell = \int_0^T \int_0^T h_\ell(\tau) \mathbf{\Psi}(t) \mathbf{\Psi}(t - \tau)^\top dt d\tau \tag{5.12}$$

In order to obtain the expression for $\mathbf{F}_\ell$, equation (5.6) was used to express the basis of user ℓ's signal space in terms of the basis of the receiver signal space. Note that all the calculations were done under the implicit assumption that the frame duration $\mathcal{T}$ is longer than the duration of all $h_\ell(t)$, $\ell = 1, \dots, L$, so that settling of all channel responses is allowed and intersymbol interference (ISI) between successive frames can be safely ignored. Equation (5.11) can also be rewritten as

$$\mathbf{y}_\ell = \mathbf{H}_\ell \mathbf{S}_\ell \mathbf{b}_\ell \tag{5.13}$$

where $\mathbf{H}_\ell = \mathbf{F}_\ell \mathcal{H}_\ell$ is the $N \times N_\ell$ channel matrix corresponding to user ℓ and incorporates, as it has already been mentioned, both the transformation between user ℓ's signal space and receiver signal space (matrix $\mathcal{H}_\ell$), and the channel $h_\ell(t)$ memory (matrix $\mathbf{F}_l$).

Therefore, after projection onto the basis function the equation of the received signal at the common receiver becomes identical to equation (5.5). In the case of a CDMA system with "time chips" spanning the signal space and processing gain N, all users and the receiver reside in the same signal space of dimension N spanned by the functions[1]

$$
\Psi_n(t) = \begin{cases} \dfrac{N}{T} & \text{for} \quad \dfrac{n-1}{N}T \le t \le \dfrac{n}{N}T \\[2mm] 0 & \text{elsewhere} \end{cases} \qquad n = 1, \ldots, N \quad (5.14)
$$

As a consequence, all matrices $\mathcal{H}_\ell = \mathbf{I}_N$ and we only have to worry about computing the matrices $\mathbf{F}_\ell$ in order to derive channel matrices.

We assume a multipath channel, with paths spaced at the chip duration and impulse response given by

$$
h_\ell(t) = \sum_{i=0}^{p_\ell-1} h_i^{(\ell)} \delta\left(t - i\frac{T}{N}\right) \qquad \ell = 1, \ldots, L \qquad (5.15)
$$

with p_ℓ being the number of paths corresponding to user ℓ and $h_i^{(\ell)}$ being the path gains. A similar channel model has been considered in [50, 51] in a discrete-time setting, and as we will see, the same channel matrices will be obtained through our approach.

The channel matrix corresponding to user ℓ is obtained from equation (5.12) as

$$
\begin{aligned}
\mathbf{F}_\ell &= \int_0^T \mathbf{\Psi}(t) \int_0^T \sum_{i=0}^{p_\ell-1} h_i^{(\ell)} \delta\left(t - i\frac{T}{N}\right) \mathbf{\Psi}(t-\tau)^\top d\tau\, dt \\[2mm]
&= \int_0^T \mathbf{\Psi}(t) \sum_{i=0}^{p_\ell-1} h_i^{(\ell)} \int_0^T \delta\left(t - i\frac{T}{N}\right) \mathbf{\Psi}(t-\tau)^\top d\tau\, dt \qquad (5.16) \\[2mm]
&= \sum_{i=0}^{p_\ell-1} h_i^{(\ell)} \int_0^T \mathbf{\Psi}(t) \mathbf{\Psi}\left(t - i\frac{T}{N}\right)^\top dt = \sum_{i=0}^{p_\ell-1} h_i^{(\ell)} \mathbf{F}_i^{(\ell)}
\end{aligned}
$$

---

[1] Normalization of "time chips" is done only to be consistent with the definition of basis functions assumed to have unit energy.

with $N \times N$ matrices $\mathbf{F}_i^{(\ell)}$ given by

$$\mathbf{F}_i^{(\ell)} = \int_0^T \mathbf{\Psi}(t) \mathbf{\Psi}\left(t - i\frac{T}{N}\right)^\top dt \tag{5.17}$$

Due to the particular form of basis functions $\mathbf{\Psi}(t)$ (non-overlapping rectangular pulses) matrices $\mathbf{F}_i^{(\ell)}$ will be sparse, with ones only in certain positions and zeros in the rest. Starting with the first term, this is equal to the identity matrix

$$\mathbf{F}_0^{(\ell)} = \int_0^T \mathbf{\Psi}(t) \mathbf{\Psi}(t)^\top dt = \mathbf{I}_N \tag{5.18}$$

since time chips do not overlap and only the product of a given function with itself is not equal to zero. As we delay the basis functions the ones along the main diagonal will be shifted downward: a delay of a chip duration T/N is equivalent with a downshift of the 1 along the corresponding column. This is due to the fact that, by delaying time chips, we will have less and less overlap between them. Thus, the second term is

$$\mathbf{F}_1^{(\ell)} = \int_0^T \mathbf{\Psi}(t) \mathbf{\Psi}\left(t - \frac{T}{N}\right)^\top dt = \begin{bmatrix} 0 & & & & \\ 1 & 0 & & & \\ 0 & 1 & \ddots & & \\ \vdots & \vdots & \ddots & 0 & \\ 0 & 0 & \cdots & 1 & 0 \end{bmatrix} \tag{5.19}$$

and so on, until the last term in which 1's will be shifted down p_ℓ positions. Hence, the channel matrix for user ℓ has the form

$$\mathbf{F}_\ell = \begin{bmatrix} h_0^{(\ell)} & 0 & 0 & \cdots & 0 & 0 & \cdots & 0 & 0 \\ h_1^{(\ell)} & h_0^{(\ell)} & 0 & \cdots & 0 & 0 & \cdots & 0 & 0 \\ \vdots & h_1^{(\ell)} & h_0^{(\ell)} & \cdots & 0 & 0 & \cdots & 0 & 0 \\ \vdots & \vdots & \ddots & \ddots & \vdots & \vdots & \vdots & \vdots & \vdots \\ h_{p_\ell - 1}^\ell & \ddots & \cdots & \ddots & h_0^{(\ell)} & 0 & \cdots & 0 & 0 \\ 0 & h_{p_\ell - 1}^{(\ell)} & \ddots & \cdots & h_1^{(\ell)} & h_0^{(\ell)} & \cdots & 0 & 0 \\ \vdots & \vdots & \ddots & \cdots & \cdots & \ddots & \ddots & \vdots & \vdots \\ 0 & 0 & 0 & \ddots & \cdots & \cdots & \ddots & h_0^{(\ell)} & 0 \\ 0 & 0 & 0 & \cdots & h_{p_\ell - 1}^\ell & \cdots & \cdots & h_1^{(\ell)} & h_0^{(\ell)} \end{bmatrix} \tag{5.20}$$

and is identical to the channel matrix used in [50, 51].

Returning to the multiaccess vector channel equation (5.2) we note that when the noise signal $\mathbf{n}$ is Gaussian, mutual information is maximized when the transmitted information signals $x_\ell(t)$ are jointly Gaussian [15, p. 405]. All that remains is to determine the exact spectral composition of each $x_\ell(t)$. We also note that the capacity region for the multiaccess vector channel defined in equation (5.2) has been established in [86]. In the same paper [86], maximization of sum capacity for the multiaccess vector channel (5.2)

$$C_s = \frac{1}{2} \log \left[\det \left(\sum_{\ell=1}^{L} \mathbf{H}_\ell \mathbf{X}_\ell \mathbf{H}_\ell^{\mathsf{T}} + \mathbf{W} \right) \right] - \frac{1}{2} \log(\det \mathbf{W}) \qquad (5.21)$$

is formulated as a convex optimization problem

$$\max_{\mathbf{X}_\ell} C_s \quad \text{subject to} \quad \text{Trace}\,[\mathbf{X}_\ell] = P_\ell, \quad \mathbf{X}_\ell \geq 0, \quad \ell = 1, \dots, L \quad (5.22)$$

and it is shown that optimal transmit covariance matrices $\mathbf{X}_\ell$, $\ell = 1, \dots, L$ satisfy a simultaneous water filling condition and can be found through an iterative water filling procedure.

2.   Sum Capacity Maximization and Interference Avoidance

Since in our approach transmit covariance matrices are expressed in terms of user codeword matrices as $\mathbf{X}_\ell = \mathbf{S}_\ell \mathbf{S}_\ell^{\mathsf{T}}$, $\ell = 1, \dots, L$ sum capacity can be written as

$$C_s = \frac{1}{2} \log \left[\det \left(\sum_{\ell=1}^{L} \mathbf{H}_\ell \mathbf{S}_\ell \mathbf{S}_\ell^{\mathsf{T}} \mathbf{H}_\ell^{\mathsf{T}} + \mathbf{W} \right) \right] - \frac{1}{2} \log(\det \mathbf{W}) \qquad (5.23)$$

Furthermore, in the context of interference avoidance we are interested in sum capacity maximization through codeword adaptation. Therefore, in our case the problem of maximizing sum capacity has a slightly different formulation than that in [86] which is mentioned in equation (5.22). More precisely, in our case we are interested in finding the codeword ensemble which maximizes sum capacity, that is

$$\max_{\mathbf{S}_\ell} C_s \quad \text{subject to} \quad \text{Trace}\left[\mathbf{S}_\ell \mathbf{S}_\ell^{\mathsf{T}}\right] = M_\ell, \quad \ell = 1, \dots, L \qquad (5.24)$$

An additional constraint in our case is given by the fact that all matrices $\mathbf{S}_\ell$ have unit norm columns.

We note that the original framework for interference avoidance presented in Chapter 2 and introduced in [55], assumes an arbitrary N-dimensional signal space for the receiver *and* all users in the multiuser system, as opposed to the multiuser system in equation (5.2) where different users are allowed to reside in different signal spaces. With a CDMA access scheme each user ℓ is assigned a finite duration unit-energy signature waveform $S_\ell(t)$, or equivalently a unit norm N-dimensional codeword s_ℓ, to convey a symbol, b_ℓ. In this setting, the received signal vector at the common receiver is described by equations (2.2) or (2.4). As it has already been noted in Chapter 2, in a colored noise background the eigen-algorithm[2] for interference avoidance converges to an optimal fixed point where the resulting codeword ensemble water fills the signal space and maximizes sum capacity defined in equation (2.14). In this case, by observing that the input covariance matrix of user ℓ is a rank one matrix [85] $X_\ell = s_\ell s_\ell^\top$, the problem of finding the optimal codeword ensemble that maximizes sum capacity in equation (2.14) can also be formulated as a spectral optimization problem. More precisely, the X_ℓ are the solution of the following optimization problem:

$$\text{maximize} \quad \frac{1}{2} \log \left[\det \left(\sum_{\ell=1}^{L} X_\ell + W \right) \right] - \frac{1}{2} \log(\det W) \qquad (5.25)$$

$$\text{subject to} \quad \begin{cases} \text{Trace}\,[X_\ell] = 1 \\ X_\ell \geq 0 \\ \text{rank}(X_\ell) = 1 \end{cases} \quad \ell = 1, \ldots, L \qquad (5.26)$$

However, because the rank constraint is not a convex constraint, the optimization problem defined in equation (5.25) subject to the constraints in equation (5.26) does not enjoy the usual global convergence properties of convex optimization problems. Thus, suboptimal fixed points of the eigen-algorithm for interference avoidance are theoretically possible, even though they have never been observed in practice when starting with a randomly chosen codeword ensemble [55], as discussed in detail in Chapter 3.

Returning to our general sum capacity maximization problem in equation (5.24) we note that this is a convex optimization problem only when each user has at least as many codewords M_ℓ as signal space dimensions N_ℓ, so that its corresponding transmit covariance matrix X_ℓ is allowed to span the whole signal space. We also note that sum capacity maximization subject to additional constraints on the rank of X_ℓ matrices is still an open research problem.

3. Greedy Interference Avoidance for Vector Channels

The fact that algorithms based on greedy interference avoidance have been shown to yield codeword ensembles that maximize sum capacity for dispersive channels – a particular case of the general multiaccess vector channel in equation (5.5) – suggests that greedy interference avoidance may also be used

---

[2]Augmented eventually with the "class warfare" procedure [54]

in the general context of multiaccess vector channels to obtain such optimal codeword ensembles. Since the second term in the sum capacity expression in equation (5.23) is fixed one needs only to maximize the determinant of the received signal covariance matrix

$$\mathbf{R} = E[\mathbf{r}\mathbf{r}^\mathsf{T}] = \sum_{\ell=1}^{L} \mathbf{H}_\ell \mathbf{S}_\ell \mathbf{S}_\ell^\mathsf{T} \mathbf{H}_\ell^\mathsf{T} + \mathbf{W} \tag{5.27}$$

In order to establish application of greedy interference avoidance for multiaccess vector channels we rewrite the received signal in equation (5.5) from the perspective of user k

$$\mathbf{r} = \mathbf{H}_k \mathbf{S}_k \mathbf{b}_k + \sum_{\ell=1,\ell\neq k}^{L} \mathbf{H}_\ell \mathbf{S}_\ell \mathbf{b}_\ell + \mathbf{n} = \mathbf{H}_k \mathbf{S}_k \mathbf{b}_k + \mathbf{z}_k \tag{5.28}$$

where $\mathbf{z}_k$ represents the interference-plus-noise seen by user k

$$\mathbf{z}_k = \sum_{\ell=1,\ell\neq k}^{L} \mathbf{H}_\ell \mathbf{S}_\ell \mathbf{b}_\ell + \mathbf{n} \tag{5.29}$$

with covariance matrix

$$\mathbf{Z}_k = E[\mathbf{z}_k \mathbf{z}_k^\mathsf{T}] = \sum_{\ell=1,\ell\neq k}^{L} \mathbf{H}_\ell \mathbf{S}_\ell \mathbf{S}_\ell^\mathsf{T} \mathbf{H}_\ell^\mathsf{T} + \mathbf{W} \tag{5.30}$$

Since $\mathbf{Z}_k$ is symmetric it can be diagonalized

$$\mathbf{Z}_k = \mathbf{E}_k \boldsymbol{\Delta}_k \mathbf{E}_k^\mathsf{T} \tag{5.31}$$

Furthermore, because $\mathbf{Z}_k$ is a positive definite covariance matrix, we can define the whitening transformation

$$\mathbf{T}_k = \boldsymbol{\Delta}_k^{-1/2} \mathbf{E}_k^\mathsf{T} \tag{5.32}$$

such that in transformed coordinates equation (5.28) is equivalent to

$$\tilde{\mathbf{r}} = \mathbf{T}_k \mathbf{r} = \mathbf{T}_k \mathbf{H}_k \mathbf{S}_k \mathbf{b}_k + \mathbf{T}_k \mathbf{z}_k = \tilde{\mathbf{H}}_k \mathbf{S}_k \mathbf{b}_k + \mathbf{w}_k \tag{5.33}$$

where $\tilde{\mathbf{H}}_k = \mathbf{T}_k \mathbf{H}_k$ is the channel matrix seen by user k in the new coordinates and $\mathbf{w}_k = \mathbf{T}_k \mathbf{z}_k$ is the equivalent "white noise" with covariance matrix $E[\mathbf{w}_k \mathbf{w}_k^\mathsf{T}] = \mathbf{T}_k \mathbf{Z}_k \mathbf{T}_k^\mathsf{T} = \mathbf{I}$ equal to the identity matrix. We note that the received signal covariance matrix in the transformed coordinates is related to the original signal covariance matrix by the equation

$$\tilde{\mathbf{R}} = E[\tilde{\mathbf{r}}\tilde{\mathbf{r}}^\mathsf{T}] = \mathbf{T}_k \mathbf{R} \mathbf{T}_k^\mathsf{T} \tag{5.34}$$

and any procedure that attempts to increase det $\tilde{\mathbf{R}}$ through adaptation of user k codewords will also increase det $\mathbf{R}$ since they are related by

$$\det \tilde{\mathbf{R}} = \det(\mathbf{T}_k \mathbf{R} \mathbf{T}_k^\top) = \det \mathbf{R} (\det \mathbf{T}_k)^2 \qquad (5.35)$$

We now apply the singular value decomposition (SVD) [66, p. 442] to the transformed channel matrix corresponding to user k

$$\tilde{\mathbf{H}}_k = \mathbf{U}_k \mathbf{D}_k \mathbf{V}_k^\top \qquad (5.36)$$

where matrix $\mathbf{U}_k$ of dimension $N \times N$ has as columns the eigenvectors of $\tilde{\mathbf{H}}_k \tilde{\mathbf{H}}_k^\top$, matrix $\mathbf{V}_k$ of dimension $N_k \times N_k$ has as columns the eigenvectors of $\tilde{\mathbf{H}}_k^\top \tilde{\mathbf{H}}_k$, and matrix $\mathbf{D}_k$ of dimension $N \times N_k$ contains the singular values of $\tilde{\mathbf{H}}_k$ on the main diagonal and zero elsewhere. We note that because $\mathbf{T}_k$ is invertible, the rank of $\tilde{\mathbf{H}}_k$ will be equal to that of $\mathbf{H}_k$. Without loss of generality we assume that $\mathbf{H}_k$ has full rank[3] N_k. Thus, the singular value matrix $\mathbf{D}_k$ can be partitioned as

$$\mathbf{D}_k = \begin{bmatrix} \tilde{\mathbf{D}}_k \\ \mathbf{0} \end{bmatrix} \qquad (5.37)$$

with $\tilde{\mathbf{D}}_k$ an $N_k \times N_k$ diagonal matrix containing the non-zero singular values along the diagonal and zeros in the rest. The left inverse of $\mathbf{D}_k$ is defined as

$$\mathbf{D}_k^\dagger = \begin{bmatrix} \tilde{\mathbf{D}}_k^{-1} & \mathbf{0} \end{bmatrix} \qquad (5.38)$$

with

$$\mathbf{D}_k^\dagger \mathbf{D}_k = \mathbf{I}_{N_k} \qquad (5.39)$$

Returning to equation (5.33) in which the SVD for transformed channel matrix $\tilde{\mathbf{H}}_k$ has been applied we have

$$\tilde{\mathbf{r}} = \mathbf{U}_k \mathbf{D}_k \mathbf{V}_k^\top \mathbf{S}_k \mathbf{b}_k + \mathbf{w}_k \qquad (5.40)$$

We pre-multiply by $\mathbf{U}_k^\top$

$$\mathbf{r}_k = \mathbf{U}_k^\top \tilde{\mathbf{r}} = \mathbf{D}_k \mathbf{V}_k^\top \mathbf{S}_k \mathbf{b}_k + \mathbf{U}_k^\top \mathbf{w}_k \qquad (5.41)$$

and define $\tilde{\mathbf{S}}_k = \mathbf{V}_k^\top \mathbf{S}_k$ and $\tilde{\mathbf{w}}_k = \mathbf{U}_k^\top \mathbf{w}_k$. Note that because both $\mathbf{U}_k$ and $\mathbf{V}_k$ are orthogonal matrices they preserve norms of vectors. Thus, columns of $\tilde{\mathbf{S}}_k$ are also unit norm as were the columns of $\mathbf{S}_k$. Also, because the equivalent

---

[3]This is not a restriction since if $\mathbf{H}_k$ is not full rank then some dimensions the user k signal space will have zero projection on the output space. Therefore we can redefine a reduced codeword matrix $\mathbf{S}_k$ which uses only dimensions with nonzero projections on the output space.

noise term $\mathbf{w}_k$ is white, then $\tilde{\mathbf{w}}_k$ will remain white with the same covariance matrix equal to the identity matrix.

$$\mathbf{r}_k = \mathbf{D}_k \tilde{\mathbf{S}}_k \mathbf{b}_k + \tilde{\mathbf{w}}_k \tag{5.42}$$

with covariance matrix given by

$$\mathbf{R}^{(k)} = E[\mathbf{r}_k \mathbf{r}_k^{\mathsf{T}}] = \mathbf{U}_k^{\mathsf{T}} \tilde{\mathbf{R}} \mathbf{U}_k \tag{5.43}$$

for which $\det \mathbf{R}^{(k)} = \det \tilde{\mathbf{R}}$ since $\mathbf{U}_k$ is an orthogonal transformation. Following a similar line of reasoning as above we note that any procedure which attempts to increase $\det \mathbf{R}^{(k)}$ through adaptation of $\tilde{\mathbf{S}}_k$ will also increase $\det \mathbf{R}$. We also note that the partitioning of the singular value matrix in equation (5.37) implies the following partition on $\mathbf{R}^{(k)}$

$$\mathbf{R}^{(k)} = \begin{bmatrix} \tilde{\mathbf{D}}_k \tilde{\mathbf{S}}_k \tilde{\mathbf{S}}_k^{\mathsf{T}} \tilde{\mathbf{D}}_k + \mathbf{I}_{N_k} & \mathbf{0} \\ \mathbf{0} & \mathbf{I}_{N-N_k} \end{bmatrix} \tag{5.44}$$

in which $\mathbf{I}_\rho$ denotes the identity matrix of order ρ and $\mathbf{0}$ denotes a matrix with all elements equal to zero.

At this point we define an equivalent problem for user k by pre-multiplying with the left inverse of $\mathbf{D}_k$ and obtain

$$\tilde{\mathbf{r}}_k = \mathbf{D}_k^{\dagger} \mathbf{r}_k = \tilde{\mathbf{S}}_k \mathbf{b}_k + \tilde{\mathbf{z}}_k \tag{5.45}$$

which is identical in form to equation (2.4) and allows straightforward application of greedy interference avoidance to optimizing user k's codeword matrix $\tilde{\mathbf{S}}_k$. The "noise" term $\tilde{\mathbf{z}}_k$ in equation (5.45) represents the interference-plus-noise from the rest of the system that is present in user k signal space and has covariance matrix

$$\tilde{\mathbf{Z}}_k = E[\tilde{\mathbf{z}}_k \tilde{\mathbf{z}}_k^{\mathsf{T}}] = \mathbf{D}_k^{\dagger} E[\tilde{\mathbf{w}}_k \tilde{\mathbf{w}}_k^{\mathsf{T}}] \mathbf{D}_k^{\dagger \mathsf{T}} = \tilde{\mathbf{D}}_k^{-2} \tag{5.46}$$

The transformed codeword matrix $\tilde{\mathbf{S}}_k$ in equation (5.45) is completely equivalent with the original codeword matrix $\mathbf{S}_k$ since they are related through an orthogonal transformation $\mathbf{V}_k^{\mathsf{T}}$. The covariance matrix of $\tilde{\mathbf{r}}_k$ is given by

$$\tilde{\mathbf{R}}^{(k)} = E[\tilde{\mathbf{r}}_k \tilde{\mathbf{r}}_k^{\mathsf{T}}] = \tilde{\mathbf{S}}_k \tilde{\mathbf{S}}_k^{\mathsf{T}} + \tilde{\mathbf{D}}_k^{-2} \tag{5.47}$$

and using the partition in equation (5.44) we have

$$\det \mathbf{R}^{(k)} = \det(\tilde{\mathbf{D}}_k \tilde{\mathbf{S}}_k \tilde{\mathbf{S}}_k^{\mathsf{T}} \tilde{\mathbf{D}}_k + \mathbf{I}_{N_k}) = \det \tilde{\mathbf{D}}_k^{-2} \det \tilde{\mathbf{R}}^{(k)} \tag{5.48}$$

which implies again that increasing $\det \tilde{\mathbf{R}}^{(k)}$ will also increase $\det \mathbf{R}^{(k)}$ which in turn implies increasing $\det \mathbf{R}$ – the determinant of the original received signal covariance matrix.

Greedy interference avoidance consists of replacing codeword corresponding to symbol m of user k by the minimum eigenvector of the autocorrelation matrix of the corresponding interference-plus-noise process in the transformed problem, that is

$$\mathbf{R}_m^{(k)} = \tilde{\mathbf{S}}_k \tilde{\mathbf{S}}_k^\top - \tilde{\mathbf{s}}_m^{(k)} \tilde{\mathbf{s}}_m^{(k)\top} + \tilde{\mathbf{D}}_k^{-2} \tag{5.49}$$

Since $\mathbf{R}^{(k)} = \mathbf{R}_m^{(k)} + \tilde{\mathbf{s}}_m^{(k)} \tilde{\mathbf{s}}_m^{(k)\top}$, our results from chapter 2 indicate that $\det \mathbf{R}^{(k)}$ and implicitly sum capacity will be monotonically increased by application of greedy interference avoidance.

We also note that, empirically we have observed that repeated application of greedy interference avoidance with various codeword replacement procedures with respect to codewords/users usually reaches a fixed point. Although we have not been able to prove this result in general, simulations have shown that, when users have at least as many codewords as signal space dimensions, this fixed point corresponds to a simultaneous water filling solution for all users in their respective signal spaces and is identical to that in [86], thus corresponding to maximum sum capacity.

4.   Algorithm Variations

Myriad interference avoidance algorithms can be formulated based on repeated application of the greedy interference avoidance procedure presented in the previous section. These are defined by the various ways in which user codewords are selected for replacement. For example, one algorithm could be defined by replacement at a given step of one codeword of a given user, followed by replacement of a randomly selected codeword a randomly selected user. Alternatively, at a given step of the algorithm, one could replace the codeword with the lowest SINR over all codewords and users. Or one could replace the codeword which will yield the maximum increase in sum capacity. One can even posit "lagged" versions of the algorithm that instead of using the minimum eigenvector, a linear superposition of the minimum eigenvector and the original codeword is used.

Nonetheless, we note that empirically we have observed that repeated application of greedy interference avoidance with various codeword replacement reaches an optimal fixed point – unless deliberately placed in a suboptimal fixed point at initialization. Unfortunately, we have been unable to prove this result in general. However, simulations have shown that when users have at least as many codewords as signal space dimensions, this fixed point reached by interference avoidance invariably corresponds to a simultaneous water filling solution for all users and must therefore be a maximum sum capacity ensemble [86].

We now formally state two algorithms for which convergence to maximum sum capacity can be proven.

The Maximum Capacity Increase Algorithm For Interference Avoidance

1 Start with a randomly chosen codeword ensemble specified by user codeword matrices $\{\mathbf{S}_k\}_{k=1}^{L}$

2 Define the equivalent problem in equation (5.45) for all users $k = 1 \ldots L$

3 Identify that codeword $\mathbf{s}_m^{(k)}$ whose replacement will maximally increase sum capacity. If no codeword will increase sum capacity, and escape methods [54] are ineffective, then STOP. Else

   (a) apply greedy interference avoidance: replace $\mathbf{s}_m^{(k)}$ by the minimum eigenvector of the corresponding $\mathbf{R}_m^{(k)}$ in equation (5.49)

   (b) return to step 2

We note that, in addition to monotonically increasing sum capacity as a consequence of applying greedy interference avoidance, the algorithm stops only if sum capacity can no longer be increased. Therefore, sum capacity must be strictly increasing. Using the same line of reasoning as in Chapters 2 and 4, it can be shown that in the limit, as the number of iterations goes to infinity, the algorithm yields codeword matrices which water fill each user's signal space, thus leading to the sum capacity maximizing simultaneous water filling [86].

The algorithm may not seem very attractive from a practical implementation point of view since it involves finding the poorest performer. However, one could imagine a stochastic update procedure where update probability increased with decreasing codeword SINR. Though not exactly the algorithm stated above, it is close enough that one could imagine similar performance. But the practical utility of this specific algorithm aside, its theoretical importance is that it is *not a water filling procedure* but still which converges to a simultaneously water-filled solution.

Of course, we can also provide a codeword update sequence which mimics iterative waterfilling:

The Eigen-Algorithm for Multiaccess Vector Channels

1 Start with a randomly chosen codeword ensemble specified by user codeword matrices $\{\mathbf{S}_k\}_{k=1}^{L}$

2 For each user $k = 1 \ldots L$

   (a) Define the equivalent problem in equation (5.45)

   (b) adjust user k codewords sequentially by applying greedy interference avoidance: the codeword corresponding to symbol m of user k is replaced by the minimum eigenvector of the corresponding $\mathbf{R}_m^{(k)}$ in equation (5.49)

(c) Iterate previous step until convergence (making use of escape methods [54] if necessary)

3 Repeat step 2.

Application of the eigen-algorithm for user k in steps 2(b)-(c) above corresponds to water filling of user k's signal space while regarding the other users as noise. The water filling process is done for all users sequentially and repeated (step 3) until convergence to a fixed point. From this perspective, iterative application of the eigen-algorithm by all users in the system is an instance of an iterative water filling procedure [86] in which each user adapts its corresponding codeword matrix, regarding all other users' signals as noise while maximizing det $\mathbf{R}$. Such a procedure is guaranteed to converge to an optimal fixed point where sum capacity is maximized.

5. Summary

We have seen that application of greedy interference avoidance monotonically increases sum capacity for general multiaccess vector channels. This result allows us to apply interference avoidance algorithms to a wide variety of scenarios including multiuser systems with multiple inputs and multiple outputs (multiuser MIMO) and asynchronous multiuser systems as will be considered explicitly in the next chapter.

Chapter 6

APPLICATIONS

The results and algorithms presented in the previous chapter are general and applicable to any communication problem for which an equivalent multiaccess vector channel representation can be obtained. We note that once the multiaccess vector channel model is determined according to the specifics of the given problem, application of interference avoidance methods becomes straightforward.

In this chapter we present application to multiple antenna systems and asynchronous CDMA systems, which are two cases of particular interest for wireless communication systems, and for which a multiaccess vector channel model is a natural choice.

1. Multiple Antenna Systems

Multiple antenna systems, used in wireless communications to provide spatial diversity, have been shown to improve system performance by mitigating the effects of multipath fading. Traditionally, spatial diversity was implemented only at one side of the communication system (mainly at the receiver). However, recent research indicates that performance can be significantly improved by using spatial diversity both at the transmitter and at the receiver.

Performance of multiple antenna systems in fading environments has been analyzed in several papers which have shown a potentially large increase in capacity. Since standard approaches are not close in performance to the theoretical limits [19, 35], new modulation schemes have been proposed and analyzed for multiple antenna systems [18, 24]. It has also been shown that presence of multipath can improve performance with an appropriate multiple antenna structure

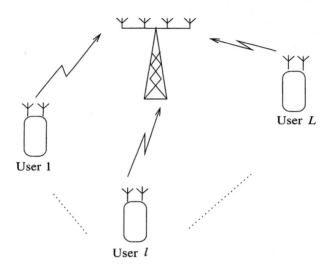

Figure 6.1.   Multiuser MIMO system in which all users and the base station are equipped with antenna arrays for transmission/reception.

[52]. Recently, an optimal power control algorithm for multiple antenna systems [87] has been proposed based on an iterative water filling scheme [86].

In this section we present application of interference avoidance methods to multiuser systems with multiple inputs and multiple outputs (MIMO) such as those associated with the uplink of a wireless system in which users and the base station are equipped with multiple antennas. As already mentioned, the approach is based on application of interference avoidance to the particular multiaccess vector channel that corresponds to a multiuser MIMO system. We note that a similar development can be found in [43]. We also note that a vector channel characterized by a channel matrix between a given transmitter and the receiver is a natural representation in the case of MIMO systems and several models can be found in the literature [18, 19, 35, 52], and that the approach is completely general and applicable to any MIMO system models regardless of the choice of signal space basis functions.

We consider a multiuser system with L users communicating with a common receiver (base station) in which all users and the base station are equipped with antenna arrays for transmission/reception. We denote by T_ℓ the number of transmit antennas corresponding to user ℓ, $\ell = 1, \ldots, L$ and by R the number of receive antennas at the base station. The system is described schematically in Figure 6.1.

Each user ℓ transmits information in frames of duration T. The channel between user ℓ's transmit antenna i and receive antenna j at the base station is characterized by the causal impulse response $h_{ij}^{(\ell)}(t)$ of duration $T_{ij}^{(\ell)}$ assumed stable (time-invariant) over the duration T of the frame. The transmitted waveform $x_i^{(\ell)}(t)$ at transmit antenna i of user ℓ convolved with this impulse response yields the corresponding waveform $y_{ij}^{(\ell)}(t)$ due to user ℓ transmit antenna i at receive antenna j

$$y_{ij}^{(\ell)}(t) = x_i^{(\ell)}(t) * h_{ij}^{(\ell)}(t) \tag{6.1}$$

The received waveform at receive antenna j is then a superposition of all such received waveforms from all transmit antennas of all users plus additive Gaussian noise

$$r_j(t) = \sum_{\ell=1}^{L} \sum_{i=1}^{N_\ell} y_{ij}^{(\ell)}(t) + n_j(t) = \sum_{\ell=1}^{L} \sum_{i=1}^{N_\ell} x_i^{(\ell)}(t) * h_{ij}^{(\ell)}(t) + n_j(t) \tag{6.2}$$

If we approached the problem using the waveform representation above, the mathematics involved would become very complex and would artificially increase its difficulty. The use of a signal space approach overcomes the difficulties associated with the waveform representation by working with equivalent signal vectors and vector channels, and making use of much simpler linear algebra methods. We note that different MIMO channel models may be obtained by using different basis functions for the signal space. For example, by using sinc functions one ends up with models similar to those in [18, 52] obtained by Nyquist sampling. Our simulation results are obtained for a MIMO channel model based on a multicarrier modulation scheme similar to that used in Chapter 4. Nevertheless, once the MIMO channel matrix $\mathbf{H}_\ell$ is obtained, application of interference avoidance methods remains unchanged.

In the equivalent signal space representation, the N-dimensional received signal vector at the base station is given by [18, 19, 52]

$$\mathbf{r} = \sum_{\ell=1}^{L} \mathbf{H}_\ell \mathbf{x}_\ell + \mathbf{n} \tag{6.3}$$

with $\mathbf{x}_\ell$ being the N_ℓ-dimensional signal vector transmitted by user ℓ, $\mathbf{H}_\ell$ the $N \times N_\ell$ MIMO channel matrix corresponding to user ℓ, and $\mathbf{n}$ the noise vector at the receiver. We note that equation (6.3) is identical in form to the general multiaccess vector channel equation (5.2), and that the dimensions of the transmitter and receiver signal spaces will depend on the number of transmit and receive antennas employed[1]. Similar to Chapter 5 users transmit information symbols as frames using a multicode CDMA approach. Thus, for a given user

---

[1] For the MIMO channel models in [18, 19, 52] these are actually equal to the number of antennas employed.

ℓ the N_ℓ-dimensional transmitted vector $\mathbf{x}_\ell$ is obtained from the sequence of symbols to be sent $\mathbf{b}_\ell = [b_1^{(\ell)} \ldots b_{M_\ell}^{(\ell)}]^\top$ through a spreading operation specified by the $N_\ell \times M_\ell$ precoding matrix

$$
\mathbf{S}_\ell = \begin{bmatrix} | & | & & | \\ \mathbf{s}_1^{(\ell)} & \mathbf{s}_2^{(\ell)} & \cdots & \mathbf{s}_{M_\ell}^{(\ell)} \\ | & | & & | \end{bmatrix} \tag{6.4}
$$

whose columns have unit norm and determine the "spreading" of corresponding symbols in the frame over the N_ℓ available dimensions. Therefore, the transmitted vector sent by user ℓ is written as $\mathbf{x}_\ell = \mathbf{S}_\ell \mathbf{b}_\ell$ which implies that the received signal vector at the base station is given by

$$
\mathbf{r} = \sum_{\ell=1}^{L} \mathbf{H}_\ell \mathbf{S}_\ell \mathbf{b}_\ell + \mathbf{n} \tag{6.5}
$$

which is identical in form to equation (5.5).

Our problem is now to optimize the precoder matrices such that the sum capacity of the multiple access vector channel defined in equation (6.5) corresponding to the multiuser MIMO system is maximized.

The fact that equation (6.5) is identical to equation (5.5) of the received signal corresponding to the multiaccess vector channel in Chapter 5 suggests that greedy interference avoidance can be directly applied to precoder optimization in the case of multiuser MIMO systems. However we note that in Chapter 5 it is assumed that all channel matrices have full rank which may not always be true for the MIMO channels, in which case some of the signal space dimensions will need to be discarded as they do not carry useful information.

Following the procedure described in Chapter 5 we rewrite the received signal from user k's perspective as in equation (5.28) and apply the whitening transformation of the interference-plus-noise seen by user k in equation (5.32) to obtain the transformed MIMO channel matrix $\tilde{\mathbf{H}}_k$ for which we obtain the SVD as in equation (5.36). We note that any vector in the N_k-dimensional input space of user k can then be represented in terms of the orthonormal set of vectors $\{\mathbf{v}_i^{(k)}\}$ representing the columns of $\mathbf{V}_k$ in equation (5.36). Similarly, any vector in the N-dimensional receiver space is representable in terms of the orthonormal set of vectors $\{\mathbf{u}_i^{(k)}\}$ representing the columns of $\mathbf{U}_k$. Furthermore, because these sets of vectors come from the SVD decomposition in equation (5.36) we can write

$$
\mathbf{v}_i^{(k)\top} \mathbf{v}_j^{(k)} = \delta_{ij} \;\Rightarrow\; \mathbf{v}_i^{(k)\top} \tilde{\mathbf{H}}_k^\top \tilde{\mathbf{H}}_k \mathbf{v}_j^{(k)} = d_i^{(k)\,2} \delta_{ij} \tag{6.6}
$$

Therefore, user k should only put energy into those vectors $\mathbf{v}_i^{(k)}$ that correspond to non-zero singular values $d_i^{(k)} \neq 0$.

Also, let us denote by ρ_k the rank of user k's transformed MIMO channel matrix $\tilde{\mathbf{H}}_k$, equal to the number of non-zero singular values. It is obvious that

$$\rho_k = \text{rank}(\tilde{\mathbf{H}}_k) \leq \min(N, N_k) \tag{6.7}$$

Then, the dimension of the column space of matrix $\tilde{\mathbf{H}}_k$ will be equal to ρ_k. Also, the dimension of the null space of $\mathbf{H}_k$ is $N_k - \rho_k$ and the dimension of the left null space is $N - \rho_k$. Because there are only ρ_k non-zero singular values and we are interested only in their corresponding eigenvectors, we can partition matrix $\mathbf{D}_k$ containing the singular values as

$$\mathbf{D}_k = \begin{bmatrix} \bar{\mathbf{D}}_k & \mathbf{0} \\ \mathbf{0} & \mathbf{0} \end{bmatrix} \tag{6.8}$$

with a $\rho_k \times \rho_k$ diagonal matrix $\bar{\mathbf{D}}_k$ which contains the nonzero singular values and zero matrices of appropriate dimensions.

We continue with the procedure as presented in Chapter 5 and obtain the equivalent received signal for user k as in equation (5.42) but in which the partition in equation (6.8) on the singular value matrix $\mathbf{D}_k$ induces the following partition of transformed precoder matrix $\tilde{\mathbf{S}}_k$

$$\tilde{\mathbf{S}}_k = \begin{bmatrix} \tilde{\mathbf{S}}_{k1} \\ \tilde{\mathbf{S}}_{k2} \end{bmatrix} \tag{6.9}$$

with $\tilde{\mathbf{S}}_{k1}$ of dimension $\rho_k \times M_k$ and $\tilde{\mathbf{S}}_{k2}$ of dimension $(N_k - \rho_k) \times M_k$.

In light of the partitions in equations (6.8) and (6.9) and because the equivalent noise term $\tilde{\mathbf{w}}_k$ is white, we can safely ignore the last $N - \rho_k$ dimensions of the received vector $\tilde{\mathbf{r}}_k$ and reduce dimensionality of the problem to the rank of the channel matrix ρ_k. This can be done because 1) no portion of user k's signal will exist in these dimensions, and 2) the noise components on these dimensions are independent of the remaining noise components. Thus, the reduced dimension problem in which only the first ρ_k components of the received vector appear can be written as

$$\bar{\mathbf{r}}_k = [\mathbf{I}_{\rho_k} \; \mathbf{0}]\tilde{\mathbf{r}}_k = \bar{\mathbf{D}}_k \bar{\mathbf{S}}_k \mathbf{b}_k + \bar{\mathbf{w}}_k \tag{6.10}$$

with $\bar{\mathbf{S}}_k = \tilde{\mathbf{S}}_{k1}$ and $\bar{\mathbf{w}} = [\mathbf{I}_{\rho_k} \; \mathbf{0}]\tilde{\mathbf{w}}_k$. The covariance matrix of the "new" noise vector is also an identity matrix of dimension ρ_k.

We are now ready for application of the greedy interference avoidance procedure, and recall that this consists of replacing one codeword m in user k transformed precoder matrix $\bar{\mathbf{S}}_k$ with the minimum eigenvector of the corresponding interference plus noise covariance matrix

$$\mathbf{R}_m^{(k)} = \bar{\mathbf{S}}_k \bar{\mathbf{S}}_k^{\mathsf{T}} - \bar{\mathbf{s}}_m^{(k)} \bar{\mathbf{s}}_m^{(k)\mathsf{T}} + \bar{\mathbf{D}}_k^{-2} \tag{6.11}$$

We note that, in addition to maximizing the SINR corresponding to codeword m of user k, greedy interference avoidance also monotonically increases sum capacity. We also note that numerous interference avoidance algorithms can be formulated based on repeated application of the greedy interference avoidance procedure, depending on the particular order in which codewords/users are selected for replacement, and that, as emphasized in previous chapter these are in general not water filling schemes although they yield a simultaneous water filling codeword ensemble.

After application of the greedy interference avoidance codeword update procedure, the full dimension updated precoder matrix is given by

$$\mathbf{S}_k = \mathbf{V}_k \begin{bmatrix} \bar{\mathbf{S}}_k \\ \mathbf{0} \end{bmatrix} \qquad (6.12)$$

which ensures that each input codeword vector is a linear combination of only those $\mathbf{v}_i^{(k)}$ which actually appear at the channel output.

A straightforward way to implement a precoder optimization algorithm for MIMO systems based on interference avoidance is to sequentially update all codewords of a given user k until convergence and then iterate this procedure for all users. This procedure defines the eigen-algorithm for multiuser MIMO systems and is formally stated below:

The Eigen-Algorithm for Multiuser MIMO Systems

1. Start with a randomly chosen set of precoder matrices $\{\mathbf{S}_\ell\}_{\ell=1}^L$

2. For each user $k = 1 \ldots L$

   (a) Compute the transformation matrix $\mathbf{T}_k$ that whitens the interference-plus-noise seen by user k

   (b) Change coordinates and compute transformed user k's MIMO channel matrix $\tilde{\mathbf{H}}_k = \mathbf{T}_k \mathbf{H}_k$

   (c) Apply SVD for $\tilde{\mathbf{H}}_k$ and project the problem onto user k's signal space

   (d) Reduce dimensionality to ρ_k the rank of the MIMO channel matrix and obtain the reduced dimension problem in equation (6.10)

   (e) Define the equivalent problem for user k by premultiplying equation (6.10) with $\bar{\mathbf{D}}_k^{-1}$

   $$\bar{\mathbf{r}}_{k,inv} = \bar{\mathbf{D}}_k^{-1} \bar{\mathbf{r}}_k = \bar{\mathbf{S}}_k \mathbf{b}_k + \bar{\mathbf{D}}_k^{-1} \bar{\mathbf{w}}_k \qquad (6.13)$$

   (f) Adjust user k's transformed precoder matrix by replacing its columns sequentially: column m of $\bar{\mathbf{S}}_k$ ($\bar{\mathbf{s}}_m^{(k)}$) is replaced by the minimum eigenvector of corresponding interference-plus-noise covariance matrix in equation (6.11)

(g) Iterate previous step until convergence (making use of escape methods [54] if the procedure stops in suboptimal points)

3 Repeat step 2.

We note that steps 2(f)-(g) represent application of the basic eigen-algorithm introduced in Chapter 2 and "water fill" user k's signal space while regarding the remaining users in the system as noise. Therefore, applied iteratively by each user, the eigen-algorithm for MIMO systems is an instance of iterative water filling and is thus guaranteed to converge to codeword ensembles which maximize sum capacity of the multiple access vector channel in equation (6.5).

However, we also note that the distributed and asynchronous nature of independent users and codeword updates might not admit such a simple tightly coordinated sequential approach. Fortunately, interference avoidance can still be applied under the assumption of asynchronous codeword updates since each update increases sum capacity. Various provably convergent flavors of the algorithm can be used, but perhaps the most satisfying feature of distributed interference avoidance is that empirically in numerical simulations convergence does not seem to depend on the specific update method employed. Thus, interference avoidance appears to be robust.

We also note, that from a practical point of view the dimensionality of the problem can be further reduced in step 2(d) of the algorithm by taking advantage of the water filling result implied by the eigen-algorithm. More precisely, the noise levels over which water filling occurs in steps 2(f)-(g) of the algorithm are given by the inverse of the non-zero singular values of the MIMO channel matrix $\tilde{\mathbf{H}}_k$, that is

$$\bar{\mathbf{D}}_k^{-2} = \begin{bmatrix} d_1^{(k)^{-2}} & & & \\ & \ddots & & \\ & & d_i^{(k)^{-2}} & \\ & & & \ddots & \\ & & & & d_{\rho_k}^{(k)^{-2}} \end{bmatrix} \tag{6.14}$$

in decreasing order of their magnitudes as yielded by the SVD

$$d_1^{(k)} \geq \ldots \geq d_i^{(k)} \geq \ldots \geq d_{\rho_k}^{(k)} \tag{6.15}$$

This implies that their inverses will be in increasing order of their magnitudes, i.e.

$$d_1^{(k)^{-2}} \leq \ldots \leq d_i^{(k)^{-2}} \leq \ldots \leq d_{\rho_k}^{(k)^{-2}} \tag{6.16}$$

Because each user k is limited to a fixed amount of transmitted power equal to Trace $\left[\mathbf{S}_k \mathbf{S}_k^\top \right] = M_k$ (coming from the fact that each user k sends M_k symbols

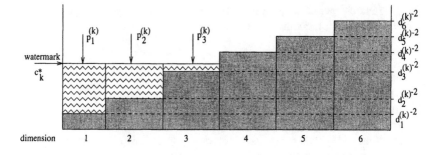

Figure 6.2. An example of water filling diagram in a signal space with 6 dimensions for which the total power M_k of user k is split only on the first 3 dimensions with minimum "noise" energy. The identities implied by water filling $c_k^* = p_1^{(k)} + d_1^{(k)^{-2}} = p_2^{(k)} + d_2^{(k)^{-2}} = p_3^{(k)} + d_3^{(k)^{-2}}$ determine $c_k^* = [M_k + (d_1^{(k)^{-2}} + d_2^{(k)^{-2}} + d_3^{(k)^{-2}})]/3 \leq d_4^{(k)^{-2}}$.

each with unit energy) we can determine how many of the ρ_k dimensions will actually carry information by looking at the "watermark" in the corresponding water filling diagram (see Figure 6.2). If we denote by n the number of dimensions water-filled by the transmitted power then the "watermark" is defined as

$$c_k^* = \frac{M_k + \sum_{i=1}^{n} d_i^{(k)^{-2}}}{n} \qquad (6.17)$$

and can be found by algorithmically by checking the following inequalities

$$\left.\begin{array}{l} c_m^{(k)^*} = \dfrac{1}{m}\left(M_k + \displaystyle\sum_{i=1}^{m} d_i^{(k)^{-2}}\right) > d_{m+1}^{(k)^{-2}} \\[2em] c_{m+1}^{(k)^*} = \dfrac{1}{m+1}\left(M_k + \displaystyle\sum_{i=1}^{m+1} d_i^{(k)^{-2}}\right) \leq d_{m+2}^{(k)^{-2}} \end{array}\right\} \implies n = m+1$$

$$(6.18)$$

The first inequality in (6.18) tells us that if we were to use only the first m dimensions to do water filling, the resulting watermark $c_m^{(k)^*}$ will be larger than the $(m+1)^{\text{st}}$ "noise level". However, the next inequality assures that when the $(m+1)^{\text{st}}$ dimension is used, the resulting watermark $c_{m+1}^{(k)^*}$ will be less than or equal to the corresponding "noise level" and therefore no additional dimensions will be water-filled.

Reducing the number of dimensions to only those which are actually water-filled by user k is advantageous from a numerical point of view since the eigen-algorithm will be performed on a lower dimensional problem. We note however, that sum capacity of the MIMO channel does not change since it is still obtained by water filling over the maximum number of dimensions with smallest "background" noise energy in user k's inverted channel signal space.

We now present numerical results from simulations and look at performance in terms of signal-to-noise ratios (SNRs) at the receiver and channel capacity. Capacity is analyzed in the context of fading channels which are characteristic of wireless communications. Due to the randomness of channel realizations, the resulting capacity is a random variable and we plot Complementary Cumulative Distribution Functions (CCDFs) similar to those in [19] from which one can see what capacity can be achieved with a given probability. More precisely, we say that an outage occurs whenever capacity is below a given value, and identify the probability of outage P_{out} from the corresponding CCDF.

The MIMO channel model used for simulations is obtained by using a multicarrier modulation scheme for transmitting information on each pair of transmit/receive antennas, which is similar to that used in Chapter 4, and for which the set of basis functions for the signal space consists of real sinusoids (sine and cosine functions). As it is customary with multicarrier modulation we assume that the frame has extended duration, $T \gg T_{ij}^{(\ell)}$, $\forall \ell$, i, j. We denote by N_c the number of frequencies which implies a $2N_c$-dimensional real signal space. The number of transmit antennas for user ℓ is T_ℓ and the number of receiver antennas is R. In this context the equivalent vector channel representation for equation (6.1) is

$$\mathbf{y}_{ij}^{(\ell)} = \mathbf{\Lambda}_{ij}^{(\ell)}\mathbf{x}_i^{(\ell)} \qquad (6.19)$$

where $\mathbf{x}_i^{(\ell)}$ and $\mathbf{y}_{ij}^{(\ell)}$ are the $2N_c$-dimensional input and output vectors, and $\mathbf{\Lambda}_{ij}^{(\ell)}$ the $2N_c \times 2N_c$ matrix containing the corresponding channel $h_{ij}^{(\ell)}(t)$ gains. The received vector at receive antenna j corresponding to the received waveform in equation (6.2) is then given by

$$\mathbf{r}_j = \sum_{\ell=1}^{L}\sum_{i=1}^{T_\ell} \mathbf{y}_{ij}^{(\ell)} + \mathbf{n}_j = \sum_{\ell=1}^{L}\sum_{i=1}^{T_\ell} \mathbf{\Lambda}_{ij}^{(\ell)}\mathbf{x}_i^{(\ell)} + \mathbf{n}_j \qquad (6.20)$$

where $\mathbf{n}_j$ is the additive noise vector at receive antenna j with covariance matrix $E[\mathbf{n}_j\mathbf{n}_j^{\top}] = \mathbf{W}_j$.

By stacking all received signal vectors from all receive antennas we can write

$$
\underbrace{\begin{bmatrix} \mathbf{r}_1 \\ \vdots \\ \mathbf{r}_j \\ \vdots \\ \mathbf{r}_R \end{bmatrix}}_{\mathbf{r}} = \sum_{\ell=1}^{L} \underbrace{\begin{bmatrix} \mathbf{\Lambda}_{11}^{(\ell)} & \cdots & \mathbf{\Lambda}_{i1}^{(\ell)} & \cdots & \mathbf{\Lambda}_{T_\ell 1}^{(\ell)} \\ \vdots & \vdots & \vdots & \vdots & \vdots \\ \mathbf{\Lambda}_{1j}^{(\ell)} & \cdots & \mathbf{\Lambda}_{ij}^{(\ell)} & \cdots & \mathbf{\Lambda}_{T_\ell j}^{(\ell)} \\ \vdots & \vdots & \vdots & \vdots & \vdots \\ \mathbf{\Lambda}_{1R}^{(\ell)} & \cdots & \mathbf{\Lambda}_{iR}^{(\ell)} & \cdots & \mathbf{\Lambda}_{T_\ell R}^{(\ell)} \end{bmatrix}}_{\mathbf{H}_\ell} \underbrace{\begin{bmatrix} \mathbf{x}_1^{(\ell)} \\ \vdots \\ \mathbf{x}_i^{(\ell)} \\ \vdots \\ \mathbf{x}_{T_\ell}^{(\ell)} \end{bmatrix}}_{\mathbf{x}_\ell} + \underbrace{\begin{bmatrix} \mathbf{n}_1 \\ \vdots \\ \mathbf{n}_j \\ \vdots \\ \mathbf{n}_R \end{bmatrix}}_{\mathbf{n}}
$$

$$(6.21)$$

which is in the form of the MIMO system equation (6.3) with $\mathbf{H}_\ell$ the $2N_c T_\ell \times 2N_c R$ matrix containing channel gain matrices for all channels in the multiple antenna link between user ℓ and the base station, $\mathbf{x}_\ell$ the $2N_c T_\ell$-dimensional transmitted vector of user ℓ, and $\mathbf{n}$ the $2N_c R$-dimensional noise vector at the base station. Under the assumption that noise vectors at different antennas are independent, then the noise covariance matrix will be block diagonal $\mathbf{W} = E[\mathbf{n}\mathbf{n}^T] = \text{diag}\{\mathbf{W}_1, \ldots, \mathbf{W}_{N_r}\}$, each block j containing the covariance of the noise that corrupts the received signal at the corresponding receive antenna $\mathbf{W}_j = E[\mathbf{n}_j\mathbf{n}_j^T]$.

The above MIMO channel model has been derived under the implicit assumption that the N_c periods of spanning sinusoids are large compared to the propagation delays between antenna elements so that the sine and cosine components are still approximately synchronized at the receiver even in the presence of multiple transmit and receive antennas. In addition, for simplicity, carrier synchronization for received signals has also been assumed.

We start with the single user multiple antenna case which was the case considered in previous work [19, 52]. For this case we analyze SNR distribution when random precoding matrices are used as well as when the optimal precoding matrices yielded by the eigen-algorithm are used, and we also look at improvement in SNR generated by an increase in number of antennas. CCDFs for sum capacity for single user as well as multiple user cases are also plotted.

1.1   Receiver SNR Distribution

In order to compute the SNRs at the receiver antennas we return to equation (6.20) which for only one user becomes

$$
\mathbf{r}_j = \sum_{i=1}^{T} \mathbf{y}_{ij} + \mathbf{n}_j = \sum_{i=1}^{T} \mathbf{\Lambda}_{ij}\mathbf{x}_i + \mathbf{n}_j \tag{6.22}
$$

Note that we have dropped the user index ℓ and replaced the number of transmit antennas by T since there is only one user in the system. The average energy of the signal at receive antenna j is given by

$$E[\mathbf{r}_j^\top \mathbf{r}_j] = \text{Trace}\left[E[\mathbf{r}_j \mathbf{r}_j^\top]\right]$$

$$= \text{Trace}\left[\sum_{p=1}^{T}\sum_{r=1}^{T}\Lambda_{pj}E[\mathbf{x}_p\mathbf{x}_r^\top]\Lambda_{rj} + \mathbf{W}_j\right] \tag{6.23}$$

$$= \text{Trace}\left[\sum_{p=1}^{T}\sum_{r=1}^{T}\Lambda_{pj}E[\mathbf{x}_p\mathbf{x}_r^\top]\Lambda_{rj}\right] + \text{Trace}\,[\mathbf{W}_j]$$

where $\mathbf{x}_p$, $\mathbf{x}_r$ represent signals coming from transmit antennas p and r respectively, which can be obtained explicitly by partitioning the $2N_cT \times M$ precoding matrix $\mathbf{S}$ in T blocks of dimensions $2N_c \times M$ stacked together

$$\mathbf{x} = \mathbf{Sb} = \begin{bmatrix} \mathbf{S}_1 \\ \vdots \\ \mathbf{S}_i \\ \vdots \\ \mathbf{S}_T \end{bmatrix} \quad \mathbf{b} = \begin{bmatrix} \mathbf{S}_1\mathbf{b} \\ \vdots \\ \mathbf{S}_i\mathbf{b} \\ \vdots \\ \mathbf{S}_T\mathbf{b} \end{bmatrix} = \begin{bmatrix} \mathbf{x}_1 \\ \vdots \\ \mathbf{x}_i \\ \vdots \\ \mathbf{x}_T \end{bmatrix} \tag{6.24}$$

With these partitions we get $E[\mathbf{x}_p\mathbf{x}_r^\top] = \mathbf{S}_p\mathbf{S}_r^\top$ and we compute the SNR at receive antenna j as the ratio of the first term in the right hand side of equation (6.23) to the second term

$$\text{SNR}_j = \frac{\text{Trace}\left[\sum_{p=1}^{T}\sum_{r=1}^{T}\Lambda_{pj}\mathbf{S}_p\mathbf{S}_r^\top\Lambda_{rj}\right]}{\text{Trace}\,[\mathbf{W}_j]} \tag{6.25}$$

Note that knowledge of channels composing the multiple antenna link implies different SNRs at different receive antennas even in the context of white noise with the same average power. This is different from [19] where the same average SNR at all receive antennas has been assumed. However, we have a similar constraint on the total transmitted power which is constant regardless of the number of transmit antennas used.

With random precoding matrices and for a particular set of channels comprising the multiple antenna link, the SNRs at different receive antennas have the bell-shaped distribution shown in Figure 6.3. The SNR at all receiver antennas has been recorded for a number of $N_c = 5$ spanning frequencies (corresponding to 10 real sinusoids), with equal number of antennas at transmitter and receiver $T = R = 1$, 2, 4, 8, and average power of the white noise $N_0 = 0.5$ at

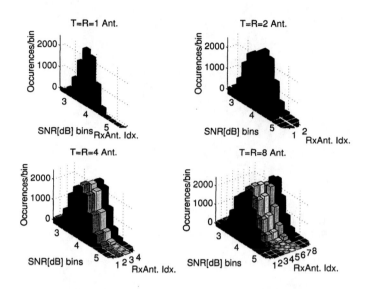

Figure 6.3.   SNR distributions for a single user multiple antenna system with random precoding matrices. Signal space dimension is $N = 10$ and average power of white noise is $N_0 = 0.5$ at each receive antenna.

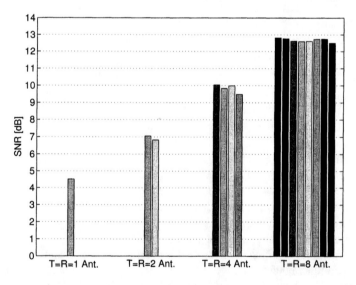

Figure 6.4.   SNRs with optimal precoding matrices yielded by the interference avoidance algorithm for a single user multiple antenna system. Signal space dimension is $N = 10$ and average power of white noise is $N_0 = 0.5$ at each receive antenna.

each receive antenna for the same set of channels but for different (random) precoding matrices. However, application of interference avoidance water fills the channels appropriately and results in a fixed set of SNRs at each receive antenna for given instances of the channel(s). As it can be seen from Figure 6.4 doubling the number of antenna elements in both transmitter and receiver results in about a 3 dB improvement in the SNR. Also note that after interference avoidance the SNR is approximately the same for all receive antennas, even though no a priori assumption about equal SNRs at each antenna [19] has been made.

1.2 Fading Channels and Outage Capacity

For the fading environments characteristic of wireless communications, the impulse responses of channels in the multiple antenna link change over time and it becomes difficult, if not impossible, to apply interference avoidance to determine optimal precoding matrices corresponding to all channel realizations that occur during the duration of the transmission. However, in such cases interference avoidance is applied using average characteristics of the channels (as in [41] and chapter 4) to determine precoder matrices which are optimal for the average channel.

We analyze the effect of fading on the sum capacity of our MIMO channel model by using precoding matrices which are optimal for the average channel (obtained by application of the eigen-algorithm) and computing the corresponding sum capacity for various realizations of the fading channel models assumed. The results are used to plot the corresponding CCDFs.

We assume the same frequency selective fading channel model as in [41] and chapter 4 which is suited for the multicarrier modulation scheme used for transmission, and which assumes flat fading of the carriers – the amplitudes of distinct carriers being scaled at the receiver by different constants. In a Rayleigh fading environment the amplitude scaling $\kappa_n^{(ij)}$ of the n-th carrier due to the channel between transmit antenna i and receive antenna j is a Rayleigh random variable with the probability density function

$$f_{\kappa_n^{(ij)}}\left(\kappa_n^{(ij)}\right) = \frac{\kappa_n^{(ij)}}{\sigma_n^{(ij)2}} e^{-\frac{\kappa_n^{(ij)2}}{2\sigma_n^{(ij)2}}} \qquad (6.26)$$

where the parameter $\sigma_n^{(\ell k)2}$ is related to the second moment of the Rayleigh random variable $E[\kappa_n^{(ij)2}] = 2\sigma_n^{(ij)2}$. The second moment of this random variable characterizes the average channel being the gain corresponding to carrier n for channel linking transmit antenna i with receive antenna j in the average channel model, i.e. $E[\kappa_n^{(ij)2}] = 2\sigma_n^{(ij)2} = \lambda_n^{(ij)}$.

For a single user system, we perform a set of experiments in which we first determine a precoding matrix which is optimal for the average channel defined in terms of a set of average values of the Rayleigh random variables, and then compute capacity values for distinct realizations of these Rayleigh random variables using equation

$$C = \frac{1}{2} \log[\det(\mathbf{HSS}^\mathsf{T}\mathbf{H}^\mathsf{T} + \mathbf{W})] - \frac{1}{2} \log(\det \mathbf{W}) \qquad (6.27)$$

where $\mathbf{H}$ is the MIMO channel matrix of the considered user. Again there is no need for user index ℓ as there is only one user in the system. The resulting set of capacity values are used to derive the CCDFs presented in Figure 6.5. In these plots we compare the case of only one transmit and one receive antenna with the case of two transmit and two receive antennas, and four transmit and four receive antennas respectively. Simulations are done with a set of $N = 5$ carrier frequencies for different average white noise power at the receiver $N_0 = 1, 0.5, 0.25, 0.1$. These values correspond to a 3 dB increase in the SNR at receiver antennas and for the average channel imply SNRs of approximately[2] 1dB, 4dB, 7dB, and 10 dB for the $T = R = 1$ antenna case, 4 dB, 7dB, 10dB and 13dB for the $T = R = 2$ antenna case, and 7 dB, 10 dB, 13 dB, and 16 dB for the $T = R = 4$ antenna case.

From the CCDFs in Figure 6.5 see that for the analyzed scenario, the use of multiple antennas at both transmitter and receiver improves outage capacity. For example, for an outage probability $P_{out} = 1\%$, capacity is increased for low SNR from less than 0.5 bits/sec/Hz to approximately 1.1 bits/sec/Hz for two transmit and receive antennas, and almost 2.7 bits/sec/Hz for four transmit and receive antennas. For high SNR, the rates increase from from about 1.5 bits/sec/Hz to about 3.6 bits/sec/Hz for two transmit and receive antennas and 5.7 bits/sec/Hz for four transmit and receive antennas.

Next we consider a multiuser MIMO system with $L = 2$ users and perform similar simulations as in the single user case with $N = 5$ carrier frequencies for different average white noise power at the receiver $N_0 = 1, 0.5, 0.25, 0.1$. Optimal precoding matrices for the average channel for two users are determined, and then sum capacity values for distinct realizations of these Rayleigh random variables is computed using equation

$$C = \frac{1}{2} \log \left[\det \left(\sum_{\ell=1}^{L} \mathbf{H}_\ell \mathbf{S}_\ell \mathbf{S}_\ell^\mathsf{T} \mathbf{H}^\mathsf{T} + \mathbf{W} \right) \right] - \frac{1}{2} \log(\det \mathbf{W}) \qquad (6.28)$$

and the resulting set of capacity values are used to derive the CCDFs presented in Figure 6.6. In these plots we compare sum capacity for the two user system in the case of only one transmit antenna per user and one receive antenna

---

[2] As it has been seen, the SNRs at receive antennas are not necessarily the same. However, when optimal codewords are used they are approximately equal.

with the case of two transmit antennas per user and two receive antennas, and four transmit antennas per user and four receive antennas respectively. From the CCDFs in Figure 6.6 we see similar improvements in sum capacity for the analyzed scenario and conclude that the use of multiple antennas in the transmitters as well as in the receiver improves outage capacity.

2.   Asynchronous CDMA Systems

An asynchronous multiuser system consists of a number users received at a common receiver with non-coincident symbol intervals. This can be a consequence of lack of synchronization for systems with users having the same data rate, or can be generated by differences in data rates for systems where distinct users are allowed to transmit at different data rates. We note that in the former case the lack of synchronization may be due to relaxing timing control which simplifies system design, while the latter case is especially interesting for future wireless systems which may have to support users with different data rates.

The capacity region of symbol-asynchronous Gaussian multiple access channels has been derived in [75] using an equivalent multiple access channel model with memory and frame synchronism [74]. Recently, for chip-based DS-CDMA systems user capacity of the asynchronous system has been analyzed and compared to that of the synchronous system [30] and a class of optimum signature sequences has been identified [70] for which there is no loss in user capacity due to asynchrony. We note that the analysis in [70] is restricted to symbol-asynchronous but chip-synchronous DS-CDMA systems, and that it does not consider the information theoretic sum capacity for the asynchronous system.

In this section we present application of interference avoidance methods to codeword optimization in an asynchronous CDMA system. Similar to [75] we consider users in the system frame synchronous and use a multicode CDMA transmission scheme. Symbols of a frame are sent in parallel using distinct signature waveforms of extended duration. This approach relaxes the synchronization requirements for the multiuser system since the frame duration will be larger than the duration of individual symbol intervals would have been if sequential transmission of symbols in a frame had been used.

An equivalent discrete time vector channel model is obtained for which application of interference avoidance is straightforward and results in codeword (or equivalently waveform) ensembles that maximize sum capacity. For this, we start by considering an asynchronous CDMA system with users transmitting at the same data rate $1/T$ for which symbol intervals corresponding to different users are not synchronized at the common receiver. Following the methodology described in [76, p. 21] one needs to introduce offsets that model the lack of alignment of symbol intervals at the receiver. As opposed to the synchronous model for which the received signal can be written by taking only one-shot of the model over the symbol interval $[0, T]$ as in equation (2.2) for the asynchronous

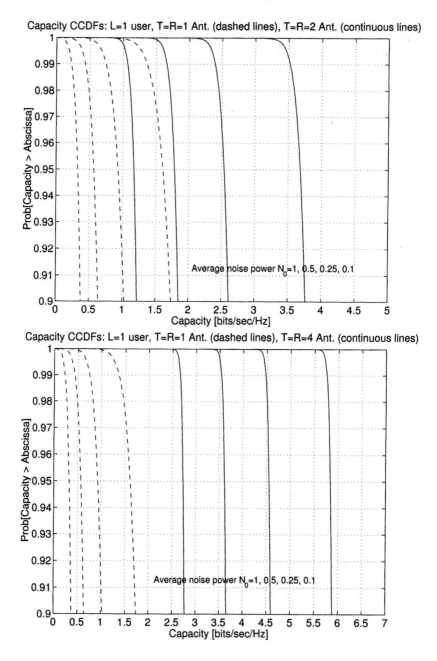

Figure 6.5.   Capacity CCDFs for single user MIMO system. The $T = R = 1$ antenna case is compared with the $T = R = 2$ antenna case (upper plot) and with the $T = R = 4$ antenna case (lower plot).

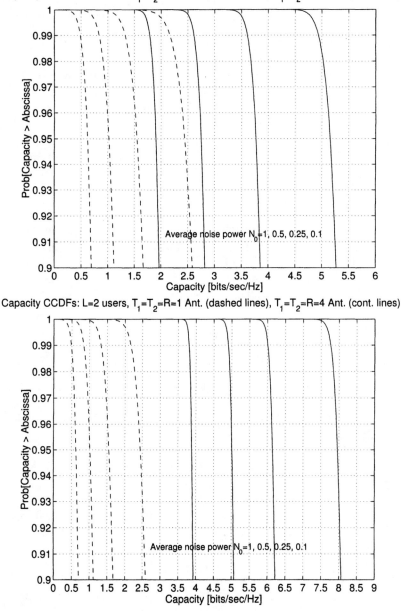

Figure 6.6. Sum capacity CCDFs for a two-user MIMO system. The $T_1 = T_2 = R = 1$ antenna case is compared with the $T_1 = T_2 = R = 2$ antenna case (upper plot) and with the $T_1 = T_2 = R = 4$ antenna case (lower plot).

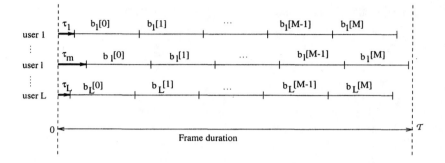

Figure 6.7.   Symbol-asynchronous users with the same data rate $1/T$ modeled as frame synchronous. Within the frame duration each user sends M symbols.

case one needs to include the offsets $\tau_\ell \in [0, T)$, $\ell = 1, \ldots, L$, and consider the fact that users send frames with many symbols $\mathbf{b}_\ell = [b_1^{(\ell)}, \ldots, b_M^{(\ell)}]$ of duration T as shown in Figure 6.7. This implies that frame-synchronism rather than symbol-synchronism is assumed, and users start and finish their transmissions within T units of each other. The received signal is then written as

$$R(t) = \sum_{\ell=1}^{L} \sum_{m=1}^{M} b_m^{(\ell)} S_\ell(t - mT - \tau_\ell) + n(t) \tag{6.29}$$

Similar to [75] we assume that the offsets τ_ℓ, $\ell = 1, \ldots, L$, are known to the receiver, but unknown to the transmitters so that they cannot advance or retard their transmissions. As noted in [75] the assumption of frame synchronism can be made at any rate with some sort of channel feedback which is an underlying assumption for systems using interference avoidance methods [55].

Since each symbol transmitted by a given user also overlaps with past/future symbols transmitted by the other users, the symbol asynchronous multiple access channel has memory and can be modeled through an equivalent multiple access channel with memory [74, 76]. Thus, one can think of approaching the asynchronous system similar to a system where symbols are subject to ISI (dispersive channels): the use of parallel transmission of information as frames through a multicode CDMA scheme using signature waveforms of extended duration to convey each symbol in the frame. While in the case of dispersive channels this approach makes ISI inconsequential, for asynchronous systems this relaxes the synchronization requirement, making it easier to be obtained in practice. Furthermore, this approach can be easily generalized to users with different data rates by considering a different number of symbols M_ℓ in a frame

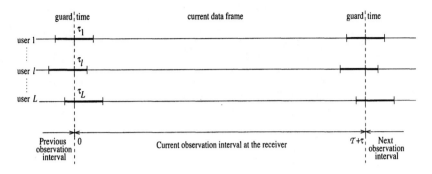

Figure 6.8. Users sending data frames of duration $\mathcal{T}$ with guard intervals.

for distinct user ℓ. The transmitted signal corresponding to user ℓ is written as

$$x_\ell(t) = \sum_{m=1}^{M_\ell} b_m^{(\ell)} s_m^{(\ell)}(t) \qquad (6.30)$$

with $s_m^{(\ell)}(t)$ being the signature waveform corresponding to symbol m of user ℓ of duration $\mathcal{T}$. We note that a similar approach has been used in [76, Ch. 4] in the analysis of multiuser detectors for asynchronous CDMA systems. The received signal at the base station is the sum of signals transmitted by all users plus additive Gaussian noise and is written as

$$R(t) = \sum_{\ell=1}^{L} x_\ell(t - \tau_\ell) + n(t) = \sum_{\ell=1}^{L} \sum_{m=1}^{M_\ell} b_m^{(\ell)} s_m^{(\ell)}(t - \tau_\ell) + n(t) \qquad (6.31)$$

Our goal is then to find an optimal ensemble of waveforms $s_m^{(\ell)}(t)$, $\ell = 1, \ldots, L$, $m = 1, \ldots, M_\ell$, for which the sum capacity of the multiple access channel is maximized. We will proceed as before, by deriving the equivalent vector channel model followed by application of greedy interference avoidance.

We assume that each user ℓ resides in a signal space of finite dimension N_ℓ implied by the frame duration $\mathcal{T}$ and finite bandwidth W_ℓ, and spanned by the vector of functions $\boldsymbol{\Psi}^{(\ell)}(t) = [\Psi_1^{(\ell)}(t) \ldots \Psi_{N_\ell}^{(\ell)}(t)]^\mathsf{T}$. Furthermore, we assume that the receiver signal space of dimension N is spanned by $\boldsymbol{\Phi}(t) = [\Phi_1(t) \ldots \Phi_N(t)]^\mathsf{T}$ and is implied by bandwidth W that includes all W_ℓ's corresponding to all users and observation interval $\mathcal{T} + \tau$ with $\tau = \max_\ell\{\tau_\ell\}$ being the largest user delay measured with respect to the beginning of the observation interval at the receiver (see Figure 6.8).

We note that, by adding the maximum delay τ to the frame duration we ensure that all waveforms corresponding to the current frame are completely observed during the observation interval at the receiver. We also note that, in order to avoid overlap between successive frames at the receiver, guard time intervals are inserted at the begining and end of each transmitted frames as it can be seen in Figure 6.8.

Each user ℓ transmits the signal $x_\ell(t)$ given by equation (6.30) which is written in terms of user ℓ basis functions as

$$x_\ell(t) = \boldsymbol{\Psi}^{(\ell)}(t)^\top \mathbf{x}_\ell = \boldsymbol{\Psi}^{(\ell)}(t)^\top \mathbf{S}_\ell \mathbf{b}_\ell \qquad (6.32)$$

with

$$\mathbf{S}_\ell = \left[\begin{array}{cccc} | & | & & | \\ \mathbf{s}_1^{(\ell)} & \mathbf{s}_2^{(\ell)} & \cdots & \mathbf{s}_{M_\ell}^{(\ell)} \\ | & | & & | \end{array} \right] \qquad \ell = 1, \ldots, L$$

the codeword matrix of user ℓ having as columns the projections of user ℓ waveforms onto the corresponding basis functions. The received signal at the common receiver contains signals of all users with corresponding delays plus additive noise

$$r(t) = \sum_{\ell=1}^{L} x_\ell(t - \tau_\ell) + n(t) \qquad (6.33)$$

and is observed over the interval $[0, \mathcal{T} + \tau]$. By projecting the received signal onto the basis functions of the receiver signal space an equivalent vector channel model is obtained for which the received signal vector is

$$\mathbf{r} = \sum_{\ell=1}^{L} \mathbf{H}_\ell \mathbf{x}_\ell + \mathbf{n} = \sum_{\ell=1}^{L} \mathbf{H}_\ell \mathbf{S}_\ell \mathbf{b}_\ell + \mathbf{n} \qquad (6.34)$$

with the $N \times N_\ell$ matrix $\mathbf{H}_\ell$ relating the user ℓ signal space and receiver signal space defined by

$$\mathbf{H}_\ell = \int_0^{\mathcal{T}+\tau} \boldsymbol{\Phi}(t) \boldsymbol{\Psi}^{(\ell)}(t - \tau_\ell)^\top dt \qquad (6.35)$$

Note that equation (6.34) is identical in form to equation (5.5), thus allowing straightforward application of greedy interference avoidance to codeword optimization as described in Chapter 5. For completeness, we formally state the generalized eigen-algorithm as it applies to codeword optimization for asynchronous systems:

The Generalized Eigen-Algorithm for Asynchronous CDMA Systems

1 Start with a randomly chosen codeword ensemble specified by the codeword matrices $\{\mathbf{S}_k\}_{k=1}^{L}$ and a specified set of user offsets $\{\tau_k\}_{k=1}^{L}$

2 Determine the set of channel matrices $\{\mathbf{H}_k\}_{k=1}^{L}$ using equation (6.35)

3 For each user $k = 1 \ldots L$

    (a) compute the transformation matrix $\mathbf{T}_k$ that whitens the interference-plus-noise seen by user k

    (b) Change coordinates and compute user k's transformed channel matrix $\tilde{\mathbf{H}}_k = \mathbf{T}_k\mathbf{H}_k$

    (c) Apply SVD for $\tilde{\mathbf{H}}_k$ and project the problem onto user k's signal space to obtain $\tilde{\mathbf{r}}_k$ in equation (5.45)

    (d) adjust user k codewords sequentially: the codeword corresponding to symbol m of user k is replaced by the minimum eigenvector of the autocorrelation matrix of the corresponding interference-plus-noise process in the inverted channel space

    (e) Iterate step (d) until convergence (making use of escape methods [54] if the procedure stops in suboptimal points)

4 Repeat step 2.

So, in summary, by using parallel transmission of symbols in a frame with signature waveforms of extended duration the asynchronous multiuser system can be modeled as a multiaccess vector channel for which application of the generalized eigen-algorithm of is straightforward. We note that, using the resulting optimal codeword ensembles implies a very simple structure at the receiver with matched filters as optimal linear receivers [78, 79].

3. Summary

We have used the results for general vector multiaccess channels from chapter 5 and applied them to two particular cases of interest in modern wireless communications – multiuser multiple input multiple output (MIMO) systems, and multiuser asynchronous systems. We also found that in a MIMO setting, uniform SINRs were achieved as a byproduct of the interference avoidance process as opposed to a constraint (or assumption) applied to the problem [19]. Thus, interference avoidance can be applied to modern systems which by various means seek to eke the maximum capacity out of limited spectral resources.

Chapter 7

EMPIRICAL STUDIES

In this chapter we present empirical studies that address some aspects of practical importance related to implementation of greedy interference avoidance algorithms. In section 1 the SINR variation during the transient phase of the greedy algorithm for interference avoidance is analyzed. We provide empirical evidence that at any step of the algorithm, the worst SINR is no worse that the minimum starting SINR. Codeword quantization issues for interference avoidance are explored in section 2. In section 3 we look at complexity issues for the eigen-algorithm and explore the reduction in receiver complexity implied by the channelization effect first noted in chapter 4 for dispersive channels. We then consider distributed implementation of interference avoidance using broadcast received covariance feedback and codeword tracking at the receiver in section 4. Finally, we make approximate analytic predictions on the performance improvement attainable through interference avoidance in a simple *multibase system* as compared to standard random binary codeword CDMA.

1. Eigen-Algorithm Transients

Consider the transient phase of the greedy interference avoidance algorithm before the codewords have settled. We ask the following question:

What if the SINR of one or several users drops below a certain level and the connection with the base station is lost?

We present empirical evidence which suggests that the minimum SINR is not decreased during practical application of the eigen-algorithm for interference avoidance. As a consequence, this ensures that if the user with the worst SINR had an acceptable connection with the base station, then usually no one's connection will be any worse than this as codewords are updated.

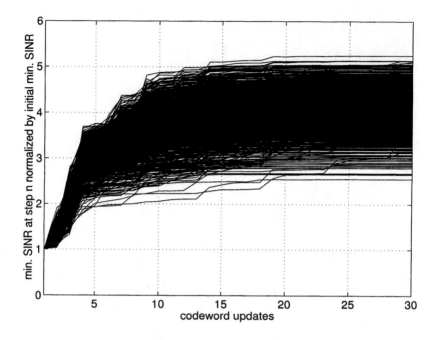

Figure 7.1. Variation of the minimum SINR normalized by the initial minimum SINR represented vs. the number of codeword updates for 1,000 random codeword ensembles for $M = 5$ users in a signal space with dimension $N = 3$ and white noise with variance $N_0 = 0.1$. We note that after 6 iterations the eigen-algorithm settles down.

We have performed experiments for various values of signal space dimensions N, and number of users M. For each value of M and N, the eigen-algorithm was applied to 1,000 random codeword ensembles and the minimum SINR value was recorded before and after a codeword update was performed. We note that convergence to near optimal codeword ensembles occurs after several iterations of the eigen-algorithm [55], and therefore each user updates its codeword several times. The minimum SINR after each codeword update was normalized by the initial minimum SINR and this value was plotted versus the number of codeword updates. From Figure 7.1, which is typical for randomly chosen initial codeword ensembles, we note that the value of the normalized minimum SINR is always larger than 1 which implies that at any step of the algorithm the minimum SINR is no lower than the initial minimum SINR. Furthermore, we note that with each codeword update the minimum SINR is increasing, and after several iterations (6 for the experiment in Figure 7.1) – once the the eigen-algorithm settles down – the minimum SINR becomes equal to the optimal value. As we have already mentioned, this empirical evidence suggests that if the user with the worst SINR has an acceptable connection with the basestation when codeword update begins, the connections of all users with the basestation will not be lost during the codeword adaptation process.

Of course there exist special cases of codeword ensembles for which updating the codeword of a given user has a negative impact on the user with the worst SINR, thus contradicting the empirical results. For example, consider $M = 5$ users, three of them optimally arranged in a subspace of dimension 2 and the other two in an orthogonal subspace of dimension 2 for a total signal space dimension of $N = 4$. We set the noise level to zero for clarity. Let the matrix containing initial user codewords be

$$
\mathbf{S} = \begin{bmatrix} 1 & \frac{1}{2} & \frac{1}{2} & 0 & 0 \\ 0 & \frac{\sqrt{3}}{2} & \frac{-\sqrt{3}}{2} & 0 & 0 \\ 0 & 0 & 0 & \sqrt{1-\epsilon^2} & \sqrt{1-\epsilon^2} \\ 0 & 0 & 0 & \epsilon & -\epsilon \end{bmatrix} \tag{7.1}
$$

The first three users reside in subspace I spanned by

$$
\mathbf{e}_1 = \begin{bmatrix} 1 \\ 0 \\ 0 \\ 0 \end{bmatrix} \quad \text{and} \quad \mathbf{e}_2 = \begin{bmatrix} 0 \\ 1 \\ 0 \\ 0 \end{bmatrix}
$$

with eigenvalues of $\frac{3}{2}$ and have SINRs of exactly 2. The last two users reside in subspace II spanned by

$$
\mathbf{e}_3 = \begin{bmatrix} 0 \\ 0 \\ 1 \\ 0 \end{bmatrix} \quad \text{and} \quad \mathbf{e}_4 = \begin{bmatrix} 0 \\ 0 \\ 0 \\ 1 \end{bmatrix}
$$

with eigenvalues $2(1 - \epsilon^2)$ and ϵ^2 respectively. The users each have SINR $\frac{1}{(1-2\epsilon^2)^2}$.

Now, consider the case where $\epsilon^2 < \frac{1}{4}$. The poorest SINRs are experienced by those users in subspace II. Now suppose a user in subspace I is chosen for codeword update. Any such user will choose $\mathbf{e}_4$ as its new codeword to obtain SINR $\frac{1}{2\epsilon^2} > 2$. However, the SINR of both original users in subspace II will *decrease* to $\frac{1}{(1-2\epsilon^2)^2+\epsilon^2}$. So, it is possible for the worst SINR to decrease upon application of interference avoidance.

Of course, one could imagine a remedy where the codeword with lowest SINR was updated at each step. To analyze this modified algorithm, let codewords $\mathbf{s}_i$ have SINR γ_i and $\gamma_1 \leq \gamma_i \; \forall i \neq 1$. Assume $\mathbf{s}_1$ is chosen and that its replacement $\mathbf{s}_1'$ has projection β_i on codeword $\mathbf{s}_i$. For now, assume that $\mathbf{s}_i \perp \mathbf{s}_1$, a constraint we will later relax.

The interference seen by s_i after replacement of s_1 is

$$\gamma_i' = \frac{1}{\frac{1}{\gamma_i} + \beta_i^2} \tag{7.2}$$

and our figure of merit is how much damage can optimal adaptation of the worst performer impose on other codewords relative the initial SINR of the worst performer. We therefore define

$$\rho = \frac{\gamma_i'}{\gamma_1} \tag{7.3}$$

as an analytic measure of such damage.

Since for a replacement to be sought we must have

$$\gamma_1' > \gamma_1 \tag{7.4}$$

and therefore

$$\frac{1}{\gamma_1} > \beta_i^2 (\frac{1}{\gamma_i} + 1) \tag{7.5}$$

since s_i sees interference $\frac{1}{\gamma_i}$ before addition of s_1' and s_i is itself interference for s_1'. However, we also note that the eigen-algorithm seeks the minimum interference dimension. So the interference seen by s_1' in the space orthogonal to s_i must be identical to that seen along s_i since otherwise the SINR could be decreased by placing all energy in the orthogonal space (if the interference level in the orthogonal space were lower), or by setting $s_1' = s_i$ if the interference along s_i were lower. Therefore we must have in total

$$\frac{1}{\gamma_1} > \frac{1}{\gamma_i} + 1 \tag{7.6}$$

and we can combine equation (7.2) with equation (7.6) to obtain

$$\rho > \frac{\frac{1}{\gamma_i} + 1}{\frac{1}{\gamma_i} + \beta_i^2} > 1 \tag{7.7}$$

since $\beta_i^2 < 1$. So, adapting the codeword with the worst performance γ_1 never results an SINR lower than γ_1 for any other codeword. q The case where s_i overlaps s_1 can be handled simply by defining $\alpha_i = s_i^T s_1$ which results in

$$\gamma_i' = \frac{1}{\frac{1}{\gamma_i} - \alpha_i^2 + \beta_i^2} \tag{7.8}$$

$$\frac{1}{\gamma_1} > \frac{1}{\gamma_i} - \alpha_i^2 + 1 \qquad (7.9)$$

and

$$\rho > \frac{\dfrac{1}{\gamma_i} - \alpha_i^2 + 1}{\dfrac{1}{\gamma_i} - \alpha_i^2 + \beta_i^2} > 1 \qquad (7.10)$$

This result makes intuitive sense since the user with worst SINR dwells in the most crowded region of the signal space. By applying the eigen-algorithm the user with worst SINR moves away from the "noisy crowd" and thereby relieves congestion. Likewise, since this worst performing user moves to the least crowded area, it cannot impose a worse SINR than it saw originally.

Finally, this result is fortunate from an algorithmic perspective since such "need based" replacement is attractive in a distributed setting. That is, one can easily imagine making the probability of codeword replacement a decreasing function of SINR.

2. Codeword Quantization

Individual user waveforms are represented as linear superpositions of orthonormal basis functions which span a signal space. For a single waveform, the real-valued superposition coefficients comprise the *codeword*. Since in a real system these values cannot be specified with infinite precision, we must ask how codeword quantization affects performance. We therefore examine the SIR obtained with optimal codewords compared to the SIR obtained with quantized optimal codewords found through interference avoidance – which differ in general from optimal quantized codewords [28].

To begin, we consider scalar quantization [49] in which the set of real numbers $\mathbb{R}$ is partitioned into $\mathcal{L}$ disjoint subsets $\{\mathcal{R}_k\}_{k=1}^{\mathcal{L}}$ and a representation point is chosen for each subset. The quantization function is

$$Q(x) = \hat{x}_k \quad \forall x \in \mathcal{R}_k \qquad (7.11)$$

and is nonlinear and noninvertible. For $\mathcal{L}$ quantization levels a number $B = \log_2 \mathcal{L}$ of bits are enough to encode them into a binary sequence[1]. Uniform and non-uniform quantization schemes are considered and their effects on the SIR and TSC will be investigated.

Simple uniform quantizers assume contiguous simply connected regions $\{\mathcal{R}_i\}_{i=2}^{\mathcal{L}} - 1$ of equal *quantization width* Δ. The design of a uniform quantizer

---

[1] $\mathcal{L}$ is chosen in general to be a power of 2.

for random variables is done to minimize the squared error distortion

$$D = E[(X - Q(X))^2] \qquad (7.12)$$

and is performed using numerical techniques in general. In contrast, for non-uniform quantizers, there is no condition that quantized regions be equal which implies fewer constraints on the minimization of distortion. Straightforward analysis results in necessary conditions for optimality, known as the Lloyd-Max conditions [49].

A set of quantization experiments has been performed using codewords obtained in 1000 trials of the eigen-algorithm. The resulting optimal codewords have been quantized using both uniform and non-uniform quantizers, with up to 32 quantization levels (which implies up to 5 bits in the representation of quantized codewords), and both the SIR and the TSC for quantized optimal codewords have been computed.

We note that, while the SIR for the optimal codeword set is the same for all users and equal to $N/(L - N)$ [55], where L is the number of users, this is no longer true for the quantized codeword set for which different users have different SIRs with a distribution which is approximately Gaussian as it can be seen from figures 7.2 and 7.3. The mean of the distribution gets closer to the optimal SIR and the variance decreases as the number of quantization levels increases.

Similar remarks can be made about the TSC value, which is equal to L^2/N for the optimal codeword set, whose resulting distributions for different number of quantization levels are plotted in figures 7.4 and 7.5.

Mean values of the SIR and TSC distributions presented in the previous plots are presented in figures 7.6 and 7.7.

Overall, we can conclude (as might be expected) that quantization degrades the performance of interference avoidance algorithms by decreasing the mean value of the SIR and increasing the mean value TSC. Furthermore, coarse quantization leads to reasonably large SIR non-uniformity across a given codeword ensemble. Each effect constitutes a degradation in overall performance. With non-uniform quantization, 16 to 32 quantization levels (corresponding to 4 – 5 bits) seem to be enough to keep the SIR distribution and TSC close to optimal values. The average number of bits required for codeword representation might be further decreased by doing entropy coding [48]. This is done by assigning codewords of variable lengths to the possible outcomes of the quantizer, such that highly probable outcomes are assigned shorter codewords, and outcomes with lower probability are assigned longer codewords.

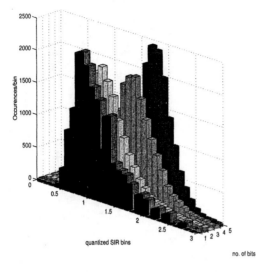

Figure 7.2. SIR distribution with uniformly quantized codewords for 1000 interference avoidance algorithm trials, $L = 15$ users with $N = 10$ dimensions.

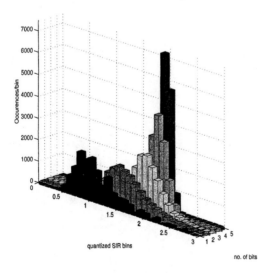

Figure 7.3. SIR distribution with non-uniformly quantized codewords for 1000 interference avoidance algorithm trials, $L = 15$ users with $N = 10$ dimensions.

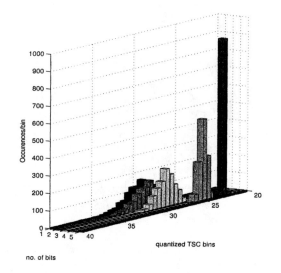

Figure 7.4.   TSC distribution with uniformly quantized codewords for 1000 interference avoidance algorithm trials, $L = 15$ users with $N = 10$ dimensions.

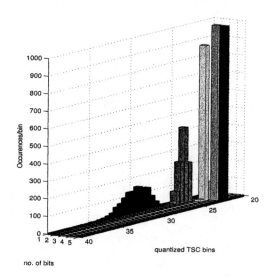

Figure 7.5.   TSC distribution with non-uniformly quantized codewords for 1000 interference avoidance algorithm trials, $L = 15$ users with $N = 10$ dimensions.

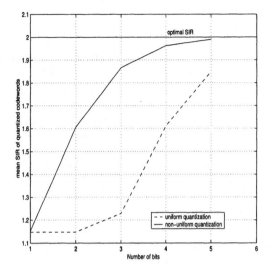

Figure 7.6. Mean values of SIR after uniform and non-uniform quantization of codewords. 1000 interference avoidance algorithm trials, $L = 15$ users, $N = 10$ dimensions.

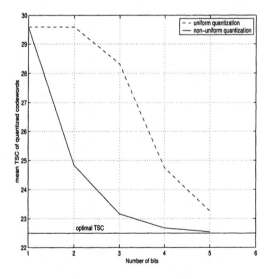

Figure 7.7. Mean values of TSC after uniform and non-uniform quantization of codewords. 1000 interference avoidance algorithm trials, $L = 15$ users, $N = 10$ dimensions.

3.   Complexity Issues

3.1   Algorithmic Complexity

At each step, the eigen-algorithm diagonalizes the correlation matrix of the interference plus noise seen by a given user, after which the given user's codeword is replaced by the minimum eigenvector. For a system with L users operating in an N-dimensional signal space, the eigen-decomposition has to be performed L times – each on an $N \times N$ matrix – for one iteration of the algorithm, with convergence occurring usually within $3 - 6$ iterations. Note that we assume the case of overloaded systems with $L \geq N$.

For symmetric matrices, computation of eigenvalues and eigenvectors can be done using various numerical techniques among which the Jacobi and Hauseholder methods [47, Ch. 11] are the most widely used. The Jacobi method consists of a sequence of orthogonal similarity transformations and its computational complexity is $O(N^3)$ (usually $18N^3$ to $30N^3$) operations when eigenvectors as well as eigenvalues are desired.

The Hauseholder method is based on reduction of the symmetric matrix to a tri-diagonal form followed by a QL algorithm (consisting of a sequence of factorizations $\mathbf{Q} \cdot \mathbf{L}$ in which $\mathbf{Q}$ is orthogonal and $\mathbf{L}$ is lower triangular). The computational complexity of this method is also $O(N^3)$, but the constants are smaller than in the case of Jacobi method (it requires $(4/3)N^3$ operations for the Hauseholder reduction to a tri-diagonal form and $3N^3$ for the QL algorithm when both eigenvalues and eigenvectors are desired).

When only a few eigenvectors[2] are desired instead of the full eigenvector set, computational complexity can be reduced by using the inverse iteration method to determine the eigenvectors. Eigenvalues will still be obtained by the Hauseholder method whose computational complexity is reduced to $(2/3)N^3 + 30N^2$ in this case. The inverse iteration method requires $O(N^2)$ operations to determine the eigenvector associated with a given eigenvalue.

Since all these methods require $O(N^3)$ operations, their use in conjunction with the eigen-algorithm for interference avoidance will imply a computational complexity of $O(LN^3)$ for the eigen-algorithm. For comparison we look at the algorithm proposed in [78], which yields codeword ensembles with the same properties as the eigen-algorithm (WBE sequence sets). This algorithm is based on a mathematical procedure of constructing a symmetric matrix with given eigenvalues and diagonal elements, and for a CDMA system with L users requires L multiplications of $L \times L$ matrices. Although in general this requires $O(L^4)$ operations, we note that the matrices involved are rotation matrices [13] and have simpler form, thus the actual complexity of the algorithm being $O(L^3)$.

---

[2]Usually less than 25% of the eigenvectors

From a different perspective, however, the eigen-algorithm is suited for distributed implementation, while the alternative algorithm in [78] is not. We note that the distributed nature of the eigen-algorithm makes it more attractive for implementation in a real system. In addition, the adaptive algorithms for estimating the eigenstructure of covariance matrices [64] or for tracking only a few eigenvectors [11] developed in the context of signal processing applications might be successfully used for further reducing the computational complexity of the eigen-algorithm as well as for more efficient implementations.

3.2   Receiver Complexity

In Chapter 4 we have seen that under reasonably loose assumptions on the channel gain matrices, two given users can overlap in at most one frequency. Such single frequency overlap can dramatically reduce the receiver complexity in a crowded system. Specifically, suppose that our signal space is spanned by N frequencies. In the real-valued formulation of Chapter 4 this implies a signal space of dimension $2N$ in which each user requires in general $2N$ filters with $2N$ coefficients each for an implied $4N^2$ multiply operations per frame per user. However, if users can only overlap in at most one frequency (2 signal space dimensions in our formulation) one might expect, depending on the actual gain matrices that each of L users will occupy on the order of $2N/L$ signal dimensions. This implies that only $2N/L$ codewords are necessary and for each codeword only $2N/L$ coefficients will be nonzero. Thus, complexity could be reduced by a factor on the order of L^2 per user receiver and L overall.

To illustrate with a more concrete example, consider $2N$ dimensions and L users. Assume that the user gain matrices are randomly perturbed identity matrices (pairwise perturbation since each sinusoid frequency must have the same gain) as given by equation (7.13) where $\epsilon_i^{(\ell)}$ is uniform random number with $|\epsilon| \leq 0.1$. Uniform white background noise is also assumed.

$$
\Lambda_\ell =
\begin{bmatrix}
1+\epsilon_1^{(\ell)} & 0 & \cdots & \cdots & \cdots & \cdots & 0 \\
0 & 1+\epsilon_1^{(\ell)} & 0 & \cdots & \cdots & \cdots & 0 \\
\vdots & 0 & \ddots & 0 & \cdots & \cdots & \vdots \\
\vdots & \cdots & 0 & \ddots & \ddots & \cdots & \vdots \\
\vdots & \cdots & \cdots & \ddots & \ddots & \ddots & \vdots \\
\vdots & \cdots & \cdots & \cdots & \ddots & 1+\epsilon_N^{(\ell)} & 0 \\
0 & \cdots & \cdots & \cdots & \cdots & 0 & 1+\epsilon_N^{(\ell)}
\end{bmatrix}
\tag{7.13}
$$

For $N = 10$ and $L = 2, 3, 4, 5, 6$ we have applied interference avoidance to a number of such randomly chosen systems and a plot of the average number of

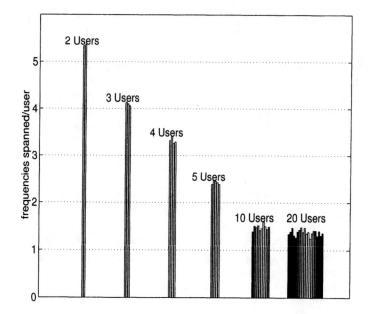

Figure 7.8. Average number of frequencies occupied per user for increasing numbers of users. The signal space is spanned by $N = 10$ frequencies and each user's gain matrix was selected according to equation (7.13).

dimensions per user is provided in Figure 7.8. As a consequence of the fact that users overlap in only one frequency we note that the more users are present in the system, the fewer frequencies are spanned by each user – with the implied decrease in user receiver complexity with L.

4.   Distributed Incremental IA

Interference avoidance could be implemented by calculating optimal codewords at the receiver and feeding these back to each transmitter independently. However, this centralized procedure can be computationally intensive at the receiver and could also overburden the required feedback channel owing to the established need of a few bits per dimension for quantization.

An alternate approach is for users to iteratively adapt their codewords in response to shared global feedback from the receiver. In turn, the receiver can adaptively track user codewords. However, greedy adaptation based on minimum eigenvector replacement could produce abrupt codeword updates which might be hard to track without disruption of associated data streams. To minimize such disruption interference avoidance schemes which allow only incre-

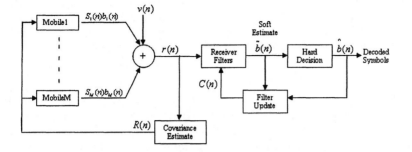

Figure 7.9. Distributed IA System Model

mental adjustments to codewords are more appropriate. These schemes, which have been introduced in Chapter 2 are:

- **Lagged IA:** codewords are adjusted in the direction of the optimal codeword.

- **Gradient Descent IA:** codewords are adjusted to most rapidly reduce the inverse SINR.

A block diagram of a system using incremental interference avoidance is shown in FIGURE 7.9, and its operation is briefly described below.

The receiver uses a separate filter for each user and produces an estimate of its transmitted symbol. During training the receiver has no *a priori* information about the transmitted codewords and starts with randomly selected filter coefficients. The transmitter sends an agreed-upon training sequence with which the receiver iteratively refines the receiver filters based on a typical error minimization criteria such as minimum mean squared error.

After training, the receiver measures the received signal covariance **R** and broadcasts it to all users periodically. Alternately, and perhaps more parsimoniously, the receiver could periodically broadcast the received vector $\mathbf{r}(n)$ and let each user construct receiver covariance estimates. We assume that each user knows its channel, so that feedback can be used by transmitters to steer transmitted codewords toward higher SINR values. The receiver decodes the symbols sent by the users and continues updating receiver filters as it did during training but now assuming that the decoded symbols are correct. This is exactly analogous to the operation of a typical adaptive equalizer [23].

Like adaptive equalization, we have found that this procedure is robust, even to certain reasonably abrupt insults such as the addition of another user to the system. One virtue of the approach is that the feedback from the receiver is global and thus does not increase with the number of users. However, from an information theoretic perspective, with some fixed number of users, the method can be less efficient than simply feeding the codewords back at discrete times [57]. Regardless, leveraging the ubiquity of adaptive equalization hardware for practical interference avoidance could be attractive. Further details about incremental interference avoidance methods can be found in [65].

5.   Comparison to Cellular

Suppose that in an interference limited system (assume no background noise) a given signal to interference ratio of γ is required for each user. We remember from chapters 2 and 3 that the maximum number of users which can be supplied the requisite γ is called the *user capacity* of the system. It is natural to ask what advantage, if any, interference avoidance can provide over existing systems such as standard synchronous CDMA with random chip-based codewords.

Since any such comparison rapidly becomes a thorny issue, even when two systems are information-theoretically identical, we will simply offer rough observations based on previous analytic work by others. A more careful treatment of the issue can be found in [55].

For a synchronous CDMA (S-CDMA) system with randomly chosen N-chip signature sequences, the user capacity is upper bounded by N/γ [69]. For a system employing interference avoidance, we have determined that at convergence, the user capacity is $N(\gamma + 1)/\gamma$. The ratio of user capacity of interference avoidance to S-CDMA systems is therefore at least

$$\eta = \gamma + 1 \tag{7.14}$$

We note that this improvement is identical to *asymptotic* results obtained for MMSE filters and random codewords [69].

For $\gamma = 1$ (0 dB) the gain is approximately 2. For $\gamma = 4$ (6 dB), the gain is approximately 5. Thus, as expected, signatures generated by interference avoidance greatly increase the single-base user capacity of S-CDMA systems. Furthermore, as outlined in previous sections, the basic receiver structure is conceptually simple – an MMSE filter and associated equalizer-like hardware for tracking and adaptation.

Of course owing to the known inefficiencies of standard CDMA for a single base, to be fair we must compare the performance of these systems for multiple bases. Once again, we will opt for simplicity of comparison. Thus, consider an S-CDMA system where each cell carries L users and the total other-cell interference energy for any given cell is $L\nu/N$ per dimension where ν is some non-negative constant. That is, we approximate all the interference energy from all other cells as white over the signal space with aggregate interference energy $L\nu$.

The SIR for each user assuming matched filters and random signatures is

$$\gamma = \frac{1}{\dfrac{L-1}{N} + \dfrac{L\nu}{N}} = \frac{N}{(\nu+1)L - 1} \qquad (7.15)$$

for a user capacity of

$$L_{\text{CDMA}} = \frac{\dfrac{N}{\gamma} + 1}{\nu + 1} = \frac{N + \gamma}{\gamma(\nu + 1)} \qquad (7.16)$$

Unfortunately, a direct comparison to a system employing interference avoidance is as yet impossible since the multiple base interference avoidance problem has not yet been solved. However, we note that for a given cell with other-cell noise energy $L\nu$, the worst interference spectral structure from a capacity standpoint would be white since interference avoidance could exploit the structure of non-white interference. Thus, taking on faith (and limited empirical observations under specific scenarios) that fixed points indeed exist for multiple base systems employing interference avoidance, we see that at each base, the "worst case" attainable ISR (interference to signal ratio) is $\frac{L+\nu L}{N} - 1$ which implies

$$L_{\text{IA}} = \frac{N(\gamma + 1)}{\gamma(\nu + 1)} \qquad (7.17)$$

– an expected improvement of $\frac{1+\gamma}{1+\frac{\gamma}{N}}$ over S-CDMA.

6.   Summary

In this chapter we have taken a look at interference avoidance from a more practical perspective. We have determined that interference avoidance does not usually hurt any other user severely. Furthermore, a simple, intuitive, and easily performed algorithm modification where the likelihood of codeword update is tied to connection quality ensures that no user's SINR ever drops below the worst SINR on the previous step.

We also considered codeword quantization and found that four to five bits per dimension were sufficient to assure reasonably uniform SIRs across codewords and near optimal ensemble performance as gauged by total square correlation.

We then examined algorithmic and receiver complexity and found that, coupled to results of chapter 4, impressive reductions in receiver complexity could be had because users would tend not to overlap even with small variations in the channel gains seen between each sender and receiver. Proceeding along these practical lines, we described incremental variants to the overall structure of interference avoidance whereby transmitters incrementally optimized their codewords in response to globally shared feedback from the receiver and the receiver adaptively tracked user codeword changes. The operations involved are essentially adaptive equalization and this bodes well for practical interference avoidance implementation owing to the ubiquity of adaptive equalizer theory and hardware in modern communications systems.

We closed with a simple comparison between interference avoidance systems and standard cellular systems which showed the potentially impressive benefits interference avoidance can confer.

Chapter 8

CONCLUSIONS

We have tried to provide a comprehensive theoretical analysis for application of interference avoidance methods to wireless communication systems. *Waveform agile* transmitters and receivers, as it is the case with software radios [1, 36, 61, 62], are a necessary prerequisite for interference avoidance which provides algorithms by which individual users in a wireless system adjust transmitted waveforms and corresponding receiver filters to better suit their operating environments.

The work is couched in a general signal space framework which allows application of interference avoidance algorithms to a wide variety of wireless communication scenarios with multiple users accessing the same communication resources. Spectrum is shared through a general CDMA scheme for transmission of information in which distinct waveforms are assigned to distinct symbols/users, but it is codewords rather than waveforms that are optimized through interference avoidance within the signal space framework.

From a user perspective, interference avoidance provides greedy procedures for SINR maximization through codeword adaptation based on feedback about interference conditions at the receiver. We have shown that rather than leading to chaos and poor performance, such selfish actions of SINR maximization lead to optimized use of the shared medium as can be seen by convergence to optimal values of global criteria such as information theoretic sum capacity or total squared correlation.

After presentation of several algorithms for interference avoidance in Chapter 2 we have focused on greedy interference avoidance in subsequent chapters. This is the procedure that defines the eigen-algorithm for interference avoidance in which the codeword of a given symbol/user is replaced by the minimum eigenvector of the corresponding interference-plus-noise covariance matrix. We have shown that greedy interference avoidance monotonically increases sum capacity, and have proved convergence of the eigen-algorithm to codeword ensembles which maximize sum capacity. We have also proved convergence of various extensions of the eigen-algorithm for dispersive and vector channels to sum capacity maximizing codeword ensembles. We note that, while such optimal codeword ensembles satisfy a *simultaneous water filling* distribution [86], interference avoidance are in general different from water filling schemes, and that the resulting water filling distribution is only an emergent property of codeword adaptation algorithms based on greedy interference avoidance.

Numerous interference avoidance algorithms can be formulated based on repeated application of this greedy interference avoidance procedure, and we have not provided a general convergence proof. We note however that, empirically we have observed that these usually reach an optimal fixed point – unless deliberately placed in a suboptimal fixed point at initialization – which suggests that greedy interference avoidance is robust with respect to codeword/user order.

The use of optimal codeword ensembles which maximize sum capacity offers potential advantages for wireless communication systems. Most notable are the fact that they provide each user a uniform SINR for all its symbols, and that the optimal linear multiuser detector for each symbol is a matched filter. Such uniform identical receiver structures may be good candidates for integration. Furthermore, when multiple users with different channels are present, as it is the case with the uplink of a wireless system, the optimal codeword ensemble exhibits natural segregation of users in signal space which may lead to potentially large reductions in receiver complexity.

From this perspective, interference avoidance provides a class of simple and effective algorithms for obtaining sum capacity maximizing codeword ensembles. Moreover, generalization of greedy interference avoidance and its application to multiaccess vector channels presented in Chapter 5 make it a versatile tool to be used in a wide range of wireless communication scenarios in which the underlying model is a multiaccess vector channel. Chapter 6 illustrates this by presenting application of greedy interference avoidance to systems with multiple inputs and multiple outputs (MIMO) and asynchronous CDMA systems.

We have also included in our monograph a presentation of issues that are of practical importance and relate to implementation of greedy interference avoidance algorithms. Empirical studies in Chapter 7 include an analysis of the transient phase of the greedy algorithm for interference avoidance, codeword quantization, complexity analysis, and a crude comparison with standard random binary CDMA systems.

We conclude our presentation with a list of theoretical and practical issues that should be further investigated in order to make interference avoidance a useful tool in the design of future wireless communication systems. We note that some of this issues are already under investigation as of this writing.

With power being a main factor in determining interference in a wireless system, incorporating some power control mechanism along with interference avoidance could provide better performance and possibly minimize user power. Very recent work [44] presents such an algorithm in which users adjusts both codeword and power so as to achieve a target SINR with minimum power, but a more thorough analysis of fixed-point properties is necessary. Convergence results for the algorithm presented in [44] or alternative interference avoidance and power control algorithms need also be established.

To determine the utility for real systems, application of interference avoidance methods in a cellular system must also be investigated. Cellular systems consist of a collection of base stations and associated users in which all users interfere with one another to some extent, and experimental results [55] have shown unstable behavior of interference avoidance algorithms when applied directly. More recent results [45, 46] have established application of interference avoidance in a collaborative scenario in which information received at all bases is centrally processed. However, the general problem of decoding one base while interfering with reception at other bases is an instance of a still mostly open information theory problem – the interference channel [3, 7, 8, 9, 21, 58, 59, 60, 63]. We think that a better understanding of the interference channel along with results on interference avoidance and iterative water filling for multiaccess vector channels may provide a starting point towards establishing application of interference avoidance in cellular systems.

Finally, for implementation purposes on a real software radio, the aspects in Chapter 7 and [67] must be complemented with more specific details and a software radio architecture for interference avoidance should be defined, as it has been done for related applications [61, 62].

We are confident, owing to the simplicity of interference avoidance concepts and advances in hardware capabilities, that interference avoidance methods will develop into practical, affordable, and efficient methods of wireless systems resource allocation.

References

[1] Special issue on software radio. *IEEE Personal Communications Magazine*, 6(4), August 1999. Editors: K-C. Chen and R. Prasad and H.V. Poor.

[2] Special issue on MIMO systems and applications. *IEEE Journal on Selected Areas in Communications*, 21(3), April 2003. Editors: M. Shafi, D. Gesbert, D. Shiu, P. J. Smith, and W. H. Tranter.

[3] R. Ahlswede. The Capacity Region of a Channel with Two Senders and Two Receivers. *Annals of Probability*, 2(5):805–814, October 1974.

[4] P. Anigstein and V. Anantharam. Ensuring convergence of the MMSE iteration for interference avoidance to the global optimum. *IEEE Transactions on Information Theory*, 49(4):873–885, April 2003.

[5] J. A. C. Bingham. Multicarrier Modulation for Data Transmission: An Idea Whose Time Has Come. *IEEE Communications Magazine*, 28(5):5–14, May 1990.

[6] R. E. Blahut. *Principles and Practice of Information Theory*. Addison-Wesley, Reading, MA, 1987.

[7] A. B. Carleial. A Case Where Interference Does Not Reduce Capacity. *IEEE Transactions on Information Theory*, 21(6):569–570, September 1975.

[8] A. B. Carleial. Interference Channels. *IEEE Transactions on Information Theory*, 24(1):60–70, January 1978.

[9] A. B. Carleial. Outer Bounds on the Capacity of Interference Channels. *IEEE Transactions on Information Theory*, 29(4):602–60, July 1983.

[10] R. S. Cheng and S. Verdú. Gaussian Multiacess Channels with ISI: Capacity Region and Multiuser Water-Filling. *IEEE Transactions on Information Theory*, 39(3):773–785, May 1993.

[11] P. Comon and G. H. Golub. Tracking a Few Extreme Singular Values and Vectors in Signal Processing. *Proceedings of the IEEE*, 78(8):1327 – 1343, August 1990.

[12] J. I. Concha and S. Ulukus. Optimization of CDMA Signature Sequences in Multipath Channels. In *Proceedings 53rd IEEE Vehicular Technology Conference – VTC'01*, volume 3, pages 1227–1239, Rhodes, Greece, May 2001.

[13] P. Cotae. An Algorithm for Obtaining Welch Bound Equality Sequences for S-CDMA Systems. *AEÜ, International Journal of Electronics and Communincations*, 55(2):95–99, 2001.

[14] P. Cotae. On Minimizing Total Weighted Squared Correlation of CDMA Systems. In *Proceedings 2003 IEEE International Symposium on Information Theory - ISIT 2003*, page 389, Yokohama, Japan, June 2003.

[15] T. M. Cover and J. A. Thomas. *Elements of Information Theory*. Wiley-Interscience, New York, NY, 1991.

[16] S. N. Diggavi. Multiuser DMT: A Multiple Access Modulation Scheme. In *Proceedings 1996 IEEE Global Telecommunications Conference - GLOBECOM '96*, pages 1566 – 1570.

[17] S. N. Diggavi. Properties of Sum-Capacity Achieving Solutions for Multiuser DMT. Private communication, October 2000.

[18] S. N. Diggavi. On Achievable Performance of Spatial Diversity Fading Channels. *IEEE Transactions on Information Theory*, 47(1):308 – 325, January 2001.

[19] G. J. Foschini and M. J. Gans. On Limits of Wireless Communications in a Fading Environment Using Multiple Antennas. *Wireless Personal Communications*, 6(3):311 – 335, March 1998.

[20] R. G. Gallager. *Information Theory and Reliable Communication*. Wiley, New York, NY, 1968.

[21] T. S. Han and K. Kobayashi. A New Achievable Rate Region for the Interference Channel. *IEEE Transactions on Information Theory*, 27(1):49 – 60, January 1981.

[22] H. Hashemi. The Indoor Radio Propagation Channel. *Proceedings of the IEEE*, 81(7):943 – 968, July 1993.

[23] S. Haykin. *Communication Systems*. Wiley, New York, NY, fourth edition, 2001.

[24] B. M. Hochwald and T. L. Marzetta. Unitary Space-Time Modulation for Multiple-Antenna Communications in Rayleigh Flat Fading. *IEEE Transactions on Information Theory*, 46(2):543 – 564, March 2000.

[25] J. L. Holsinger. Digital Communication Over Fixed Time-Continuous Channels With Memory - With Special Application to Telephone Channels. Technical Report 366, MIT - Lincoln Lab., 1964.

[26] M. L. Honig, K. Steiglitz, and S. A. Norman. Optimization of Signal Sets for Partial-Response Response Channels - Part I: Numerical Techniques. *IEEE Transactions on Information Theory*, 37(5):1327–1341, September 1991.

[27] R. A. Horn and C. A. Johnson. *Matrix Analysis*. Cambridge University Press, Cambridge, United Kingdom, 1985.

[28] G. N. Karystinos and D. A. Pados. New Bounds on the Total Squared Correlation and Optimum Design of DS-CDMA Binary Signature Sets. *IEEE Transactions on Communications*, 51(1):48–51, January 2003.

[29] S. Kasturia, J. T. Aslanis, and J. M. Cioffi. Vector Coding For Partial Response Channels. *IEEE Transactions on Information Theory*, 36(4):741–762, July 1990.

[30] Kiran and D. Tse. Effective Interference and Effective Bandwith of Linear Multiuser Receivers in Asynchronous CDMA Systems. *IEEE Transactions on Information Theory*, 46(4):1426 – 1447, July 2000.

[31] H. J. Landau and H. O. Pollack. Prolate Spheroidal Wave Functions, Fourier Analysis and Uncertainty – III: The Dimension of the Space of Essentially Time- and Band-Limited Signals. *The Bell System Technical Journal*, 41(4):1295–1335, July 1962.

[32] J. W. Lechleider. The Optimum Combination of Block Codes and Receivers for Arbitrary Channels. *IEEE Transactions on Communications*, 38(3):615–621, May 1990.

[33] U. Madhow and M. L. Honig. MMSE Interference Suppression for Direct-Sequence Spread-Spectrum CDMA. *IEEE Transactions on Communications*, 42(12):3178–3188, December 1994.

[34] A. W. Marshall and I. Olkin. *Inequalities: Theory of Majorization and its Applications*. Academic Press, Orlando, FL, 1979.

[35] T. L. Marzetta and B. M. Hochwald. Capacity of a Mobile Multiple-Antenna Communication Link in Rayleigh Flat Fading. *IEEE Transactions on Information Theory*, 45(1):139 – 157, January 1999.

[36] J. Mitola. The Software Radio Architecture. *IEEE Communications Magazine*, 33(5):26–38, May 1995.

[37] R. Negi and J. Cioffi. Pilot Tone Selection for Channel Estimation in a Mobile OFDM System. *IEEE Transactions on Consumer Electronics*, 44(3):1122–1128, August 1998.

[38] S. Ohno, P. Anghel, G. Giannakis, and Z. Luo. Multicarrier Multiple Access is Sum-Rate Optimal for Block Transmissions over Circulant ISI Channels. In *Proceedings 2002 IEEE International Conference on Communications – ICC'02*, volume 3, pages 1656–1660, May 2002.

[39] A. Papoulis. *Probabily, Random Variables, and Stochastic Processes*. McGraw-Hill, Boston, MA, third edition, 1991.

[40] D. C. Popescu, O. Popescu, and C. Rose. Interference Avoidance for Multiaccess Vector Channels. In *Proceedings 2002 IEEE International Symposium on Information Theory - ISIT'02*, page 499, Lausanne, Switzerland, July 2002.

[41] D. C. Popescu and C. Rose. Fading Channels and Interference Avoidance. In *Proceedings 39th Allerton Conference on Communication, Control, and Computing*, pages 1073–1074, Monticello, IL, October 2001.

[42] D. C. Popescu and C. Rose. Interference Avoidance Applied to Multiaccess Dispersive Channels. In *Proceedings 35th Annual Asilomar Conference on Signals, Systems, and Computers*, volume II, pages 1200–1204, Pacific Grove, CA, November 2001.

[43] D. C. Popescu and C. Rose. Interference Avoidance and Multiuser MIMO Systems. *International Journal of Satellite Communications and Networking*, 21(1):143–161, January 2003.

[44] D. C. Popescu and C. Rose. Interference Avoidance and Power Control for Uplink CDMA Systems. In *Proceedings 2003 IEEE Vehicular Technology Conference - VTC 2003 Fall*, Orlando, FL, October 2003.

[45] O. Popescu and C. Rose. Interference Avoidance and Sum Capacity for Multibase Systems. In *Proceedings 39th Allerton Conference on Communication, Control, and Computing*, pages 1036–1045, Monticello, IL, October 2001.

[46] O. Popescu and C. Rose. Minimizing Total Squared Correlation with Multiple Receivers. In *Proceedings 39th Allerton Conference on Communication, Control, and Computing*, pages 1063–1072, Monticello, IL, October 2001.

[47] W. H. Press, B. P. Flannery, S. A. Teukolsky, and W. T. Vetterling. *Numerical Recipes. The Art of Sceintific Computing*. Cambridge University Press, Cambridge, United Kingdom, first edition, 1988.

[48] J. G. Proakis. *Digital Communications*. McGraw Hill, Boston, MA, fourth edition, 2000.

[49] J. G. Proakis and M. Salehi. *Communication Systems Engineering*. Prentice Hall, Englewood Cliffs, NJ, 1994.

[50] G. S. Rajappan and M. L. Honig. Spreading Code Adaptation for DS-CDMA with Multipath. In *Proceedings 2000 IEEE Military Communications Conference – MILCOM 2000*, volume 2, pages 1164–1168, Los Angeles, CA, October 2000.

[51] G. S. Rajappan and M. L. Honig. Signature Sequence Adaptation for DS-CDMA with Multipath. *IEEE Journal on Selected Areas in Communications*, 20(2):384–395, February 2002.

[52] G. G. Raleigh and J. M. Cioffi. Spatio-Temporal Coding for Wireless Communication. *IEEE Transactions on Communications*, 46(3):357–366, March 1998.

[53] P. B. Rapajic and B. S. Vucetic. Linear Adaptive Transmitter-Receiver Structures for Asynchronous CDMA Systems. *European Transactions on Telecommunications*, 6(1):21 – 27, Jan. - Feb. 1995.

[54] C. Rose. CDMA Codeword Optimization: Interference Avoidance and Convergence Via Class Warfare. *IEEE Transactions on Information Theory*, 47(6):2368–2382, September 2001.

[55] C. Rose, S. Ulukus, and R. Yates. Wireless Systems and Interference Avoidance. *IEEE Transactions on Wireless Communications*, 1(3):415–428, July 2002.

[56] M. Rupf and J.L. Massey. Optimum Sequence Multisets for Synchronous Code-Division Multiple-Access Channels. *IEEE Transactions on Information Theory*, 40(4):1226–1266, July 1994.

[57] W. Santipach and M. L. Honig. Signature Optimization for DS-CDMA with Limited Feedback. In *Proceedings 7th IEEE International Symposium on Spread Spectrum Techniques and Applications - ISSSTA'02*, volume 1, pages 180–184, Prague, Czech Republic, September 2002.

[58] H. Sato. Two-User Communication Channels. *IEEE Transactions on Information Theory*, 23(3):295–304, May 1977.

[59] H. Sato. On the Capacity Region of a Discrete Two-User Channel for Strong Interference. *IEEE Transactions on Information Theory*, 24(3):377–779, May 1978.

[60] H. Sato. The Capacity of the Gaussian Interference Channel Under Strong Interference. *IEEE Transactions on Information Theory*, 27(6):786–788, November 1981.

[61] I. Seskar and N. Mandayam. A Software Radio Architecture for Linear Multiuser Detection. *IEEE Journal on Selected Areas in Communications*, 17(5):814 – 823, May 1999.

[62] I. Seskar and N. Mandayam. Software Defined Radio Architectures for Interference Cancellation in DS-CDMA Systems. *IEEE Personal Communications Magazine*, 6(4):26 – 34, August 1999.

[63] C. E. Shannon. Two-way Communication Channels. In *Proceedings 4th Berkeley Symposium on Mathematics, Statistics, and Probabilities*, pages 611–644. University of California Press, Berkeley, 1961.

[64] K. C. Sharman. Adaptive Algorithms for Estimating the Complete Covariance Eigenstructure. In *Proceedings 1986 International Conference on Acoustics, Speech, and Signal Processing - ICASSP '86*, pages 27.15.1 – 27.15.4, Tokyo, Japan, 1986.

[65] J. Singh and C. Rose. Distributed Incremental Interference Avoidance. In *Proceedings 2003 IEEE Global Telecommunications Conference - GLOBECOM '03*, San Francisco, CA, December 2003.

[66] G. Strang. *Linear Algebra and Its Applications*. Harcourt Brace Jovanovich College Publishers, San Diego, CA, third edition, 1988.

[67] D. Tabora. An Analysis of Covariance Estimation, Codeword Feedback, and Multiple Base Performance of Interference Avoidance. Master's thesis, Rutgers University, Department of Electrical and Computer Engineering, 2001. Thesis Director: Prof. C. Rose.

[68] D. Tse and S. Hanly. Multi-access Fading Channels. Part I: Polymatroid Structure, Optimal Resource Allocation and Throughput Capacities. *IEEE Transactions on Information Theory*, 44(7):2796 – 2815, November 1998.

[69] D. Tse and S. Hanly. Linear Multiuser Receivers: Effective Interference, Effective Bandwidth and User Capacity. *IEEE Transactions on Information Theory*, 45:641–657, March 1999.

[70] S. Ulukus and R. Yates. Optimum Signature Sequence Sets for Asynchronous CDMA Systems. In *Proceedings 38th Allerton Conference on Communication, Control, and Computing*, volume I, pages 307–316, October 2000.

[71] S. Ulukus and R. Yates. Iterative Construction of Optimum Signature Sequence Sets in Synchronous CDMA Systems. *IEEE Transactions on Information Theory*, 47(5):1989–1998, July 2001.

[72] H. L. Van Trees. *Detection, Estimation, and Modulation Theory, Part I*. Wiley, New York, NY, 1968.

[73] S. Verdú. Capacity Region of Gaussian CDMA Channels. The Symbol-Synchronous Case. In *Proceedings 24th Allerton Conference on Communication, Control, and Computing*, pages 1025–1034, October 1986.

[74] S. Verdú. Multiple-Access Channels with Memory with and without Frame Synchronism. *IEEE Transactions on Information Theory*, 35(3):605–619, May 1989.

[75] S. Verdú. The Capacity Region of the Symbol-Asynchronous Gaussian Multiple-Access Channel. *IEEE Transactions on Information Theory*, 35(4):733–751, July 1989.

[76] S. Verdú. *Multiuser Detection*. Cambridge University Press, Cambridge, United Kingdom, 1998.

[77] M. Vetterli and J. Kovacević. *Wavelets and Subband Coding*. Prentice Hall, Englewood Cliffs, NJ, 1995.

[78] P. Viswanath and V. Anantharam. Optimal Sequences and Sum Capacity of Synchronous CDMA Systems. *IEEE Transactions on Information Theory*, 45(6):1984–1991, September 1999.

[79] P. Viswanath, V. Anantharam, and D. Tse. Optimal Sequences, Power Control and Capacity of Spread Spectrum Systems with Multiuser Linear Receivers. *IEEE Transactions on Information Theory*, 45(6):1968–1983, September 1999.

[80] P. Viswanath, D. Tse, and V. Anantharam. Asymptotically Optimal Waterfilling in Vector Multiple Access Channels. *IEEE Transactions on Information Theory*, 47(1):241 – 267, January 2001.

[81] P. Viswanath and Anantharam V. Total Capacity of Multiaccess Vector Channels. Technical Memorandum M99/47, Electronics Research Laboratory, University of California, Berkeley, 1999.

[82] L. R. Welch. Lower Bounds on the Maximum Cross Correlation of Signals. *IEEE Transactions on Information Theory*, IT-20(3):397–399, May 1974.

[83] X. G. Xia. *Modulated Coding for Intersymbol Interference Channels*. Marcel Dekker, New York, NY, 2000.

[84] N. Yee and J. P. Linnartz. Multi-Carrier CDMA in an Indoor Wireless Radio Channel. Technical Memorandum M94/6, Electronics Research Laboratory, University of California, Berkeley, 1994.

[85] W. Yu. Interference Avoidance and Iterative Water Filling: A Connection. Private communication, May 2001.

[86] W. Yu, W. Rhee, S. Boyd, and J. M. Cioffi. Iterative Water-Filling for Gaussian Vector Multiple Access Channels. In *Proceedings 2001 IEEE International Symposium on Information Theory - ISIT'01*, page 322, Washington, DC, June 2001. Submitted for journal publication.

[87] W. Yu, W. Rhee, and J. M. Cioffi. Optimal Power Control in Multiple Access Fading Channels with Multiple Antennas. In *Proceedings 2001 IEEE International Conference on Communications – ICC'01*, volume 2, pages 575–579, May 2001.

Index

133